The Politics of Education and Technology

Palgrave Macmillan's Digital Education and Learning

Much has been written during the first decade of the new millennium about the potential of digital technologies to produce a transformation of education. Digital technologies are portrayed as tools that will enhance learner collaboration and motivation and develop new multimodal literacy skills. Accompanying this has been the move from understanding literacy on the cognitive level to an appreciation of the sociocultural forces shaping learner development. Responding to these claims, the Digital Education and Learning Series explores the pedagogical potential and realities of digital technologies in a wide range of disciplinary contexts across the educational spectrum both in and outside of class. Focusing on local and global perspectives, the series responds to the shifting landscape of education, the way digital technologies are being used in different educational and cultural contexts, and examines the differences that lie behind the generalizations of the digital age. Incorporating cutting-edge volumes with theoretical perspectives and case studies (single authored and edited collections), the series provides an accessible and valuable resource for academic researchers, teacher trainers, administrators, and students interested in interdisciplinary studies of education and new and emerging technologies.

Series Editors:

Michael Thomas is a senior lecturer at the University of Central Lancashire and editor in chief of the *International Journal of Virtual and Personal Learning Environments* (IJVPLE).

James Paul Gee is a Mary Lou Fulton Presidential Professor at Arizona State University. His most recent book is *Policy Brief: Getting over the Slump: Innovation Strategies to Promote Children's Learning* (2008).

John Palfrey is the head of school at Phillips Academy, Andover, and a senior research fellow at the Berkman Center for Internet & Society at Harvard. He is coauthor of *Born Digital: Understanding the First Generation of Digital Natives* (2008).

Digital Education
Edited by Michael Thomas

Digital Media and Learner Identity: The New Curatorship
By John Potter

Rhetoric/Composition/Play through Video Games: Reshaping Theory and Practice of Writing
Edited by Richard Colby, Matthew S. S. Johnson, and Rebekah Shultz Colby

Computer Games and Language Learning
By Mark Peterson

The Politics of Education and Technology: Conflicts, Controversies, and Connections
Edited by Neil Selwyn and Keri Facer

The Politics of Education and Technology

Conflicts, Controversies, and Connections

Edited by

*Neil Selwyn and
Keri Facer*

First published in 2013 by
PALGRAVE MACMILLAN®
in the United States—a division of St. Martin's Press LLC,
175 Fifth Avenue, New York, NY 10010.

Where this book is distributed in the UK, Europe and the rest of the world,
this is by Palgrave Macmillan, a division of Macmillan Publishers Limited,
registered in England, company number 785998, of Houndmills,
Basingstoke, Hampshire RG21 6XS.

Palgrave Macmillan is the global academic imprint of the above companies
and has companies and representatives throughout the world.

Palgrave® and Macmillan® are registered trademarks in the United States,
the United Kingdom, Europe and other countries.

ISBN: 978–1–137–03197–6

Library of Congress Cataloging-in-Publication Data

The politics of education and technology : conflicts, controversies, and
connections / Edited by Neil Selwyn & Keri Facer.
pages cm.—(Digital Education and Learning)
Includes bibliographical references.
ISBN 978–1–137–03197–6 (hardcover : alk. paper)
1. Educational technology. 2. Technological innovations—Social
aspects. 3. Education—Effect of technological innovations on. I. Selwyn,
Neil. II. Facer, Keri, 1972–

LB1028.3.P65 2013
371.33—dc23   2013010741

A catalogue record of the book is available from the British Library.

Design by Newgen Knowledge Works (P) Ltd., Chennai, India.

First edition: September 2013

10 9 8 7 6 5 4 3 2 1

Contents

Series Foreword

Much has been written during the first decade of the new millennium about the potential of digital technologies to radically transform education and learning. Typically such calls for change spring from the argument that traditional education no longer engages learners or teaches them the skills required for the twenty-first century. Digital technologies are often described as tools that will enhance collaboration and motivate learners to reengage with education and enable them to develop the new multimodal literacy skills required for today's knowledge economy. Using digital technologies is a creative experience in which learners actively engage with solving problems in authentic environments that underline their productive skills rather than merely passively consuming knowledge. Accompanying this argument has been the move from understanding literacy on the cognitive level to an appreciation of the sociocultural forces shaping learner development and the role communities play in supporting the acquisition of knowledge.

Emerging from this context the Digital Education and Learning series was founded to explore the pedagogical potential and realities of digital technologies in a wide range of disciplinary contexts across the educational spectrum around the world. Focusing on local and global perspectives, the series responds to the shifting demands and expectations of educational stakeholders, the ways new technologies are actually being used in different educational and cultural contexts, and examines the opportunities and challenges that lie behind the myths and rhetoric of digital age education. The series encourages the development of evidence-based research that is rooted in an understanding of the history of technology, as well as open to the potential of new innovation, and adopts critical perspectives on technological determinism as well as techno-skepticism.

While the potential for changing the way we learn in the digital age is significant, and new sources of information and forms of interaction have developed, many educational institutions and learning environments have changed little from those that existed over one hundred years ago. Whether in the form of smartphones, laptops, or tablets, digital technologies may be increasingly ubiquitous in a person's social life but marginal in their daily educational experience once they enter a classroom. Although many people increasingly invest more and more time on their favorite social media site, integrating these technologies into curricula or formal learning environments remains a significant challenge, if indeed it is a worthwhile aim in the first place. History tells us that change in educational contexts, if it happens at all in ways that were intended, is typically more "incremental" and rarely "revolutionary." Understanding the development of learning technologies in the context of a historically informed approach therefore is one of the core aspects of the series, as is the need to understand the increasing internation-alization of education and the way learning technologies are culturally medi-ated. While the digital world appears to be increasingly "flat," significant challenges continue to exist, and the series will problematize terms that have sought to erase cultural, pedagogical, and theoretical differences rather than understand them. "Digital natives," "digital literacy," "digital divide," "digi-tal media"—these and such mantras as "twenty-first-century learning"—are phrases that continue to be used in ways that require further clarification and critical engagement rather than unquestioning and uncritical acceptance.

The series aims to examine the complex discourse of digital technolo-gies and to understand the implications for teaching, learning, and profes-sional development. By mixing volumes with theoretical perspectives with case studies detailing actual teaching approaches, whether on or off cam-pus, in face-to-face, fully online, or blended-learning contexts, the series will examine the emergence of digital technologies from a range of new international and interdisciplinary perspectives. Incorporating original and innovative volumes with theoretical perspectives and case studies (single authored and edited collections), the series aims to provide an accessible and valuable resource for academic researchers, teacher trainers, administrators, policymakers, and learners interested in cutting-edge research on new and emerging technologies in education.

This new collection of 11 chapters edited by Neil Selwyn and Keri Facer entitled, *The Politics of Educational Technology*, is both timely and signifi-cant, and provides a much-needed critical analysis of the wider social, politi-cal, and economic contexts in which digital technologies are used today. The collection attempts to raise important issues about the future direction of research in the field and to reposition educational technology beyond its merely local concern with classroom practices and providing "new tips" about how to enhance teaching and learning. Moving beyond a predominantly psychological approach to education and raising important concerns about the widespread acceptance of social constructivism, the book calls for a more sociological perspective on educational technology. Such an approach is evident across the book's 11 chapters, which collectively unravel the assumption that the use of digital technologies in education is an apolitical narrative based on the absence of ideology or contested values. Discussing a variety of themes from one-to-one laptop programs to e-safety and mobile learning, each of the chapters underline how many of these microissues are in fact "proxy battles for much wider controversies and conflicts surrounding the nature, form and function of education in the twenty-first century—i.e. tensions between market and state; private interests and public good; and the primacy of individuals as opposed to the collective."

The book deconstructs the overarching "means-end" and "deterministic" thinking that has dominated the integration of educational technology in schools in many parts of the world. It asks a series of key questions about the potential integration of digital technologies in the future, outlining the need for a new research approach based on a "thick description" of the cultural context. This multilayered and critical approach establishes the need to look beneath the apparently value-neutral claims made in the name of digital technologies and to identify the inequalities of access, power, and skills that influence and mediate their use. It concludes by outlining a pathway between the polarized alternatives of overblown hope and disengaged skepticism, and focuses on ways to produce "engaged critical research" that is sensitively attuned to the politics of education and technology and the interests that lie behind them.

The Politics of Educational Technology is a much-needed and worthy addition to the Digital Education and Learning series, and we welcome its intervention and efforts to reposition research in the field over the next decade. The volume will be valuable for international educators, researchers, and policymakers and contribute to opening up new spaces for discussion about the myths and realities of educational technology in the digital age.

Series Editors

JAMES PAUL GEE,
Mary Lou Fulton Presidential
Professor of Literacy Studies,
Arizona State University

MICHAEL THOMAS,
Senior Lecturer in Language Learning Technologies,
University of Central Lancashire

JOHN PALFREY,
Head of School,
Phillips Academy, Andover

CHAPTER 1

Introduction: The Need for a Politics of Education and Technology

Neil Selwyn and Keri Facer

Introduction

Digital technology is now a prominent feature of education provision and practice in many countries and contexts. Mobile telephony, internet use, and other forms of computing are familiar, everyday tools for many people in developed and developing nations. Billions of personally owned digital devices are in almost continuous use, and billions of others are used communally in shared, public settings. Governments of nearly every country in the world now have well-established policy drives and programs seeking to encourage and support the use of digital technologies in schools, colleges, and universities. Digital technology is a topic that is of significance to a global educational audience.

Yet for such a significant issue, there remains relatively limited analysis of the politics, the economics, the cultures, and the ethics of digital technology in education. Academic study of educational technology, such as that found in leading journals such as *Computers and Education*, for example, has developed into a field dominated by psychological (usually social-psychological) perspectives on learning and teaching. Many educational technology researchers proudly align themselves with the "learning sciences" rather than the social sciences. This research is therefore often concerned primarily with matters relating to individual behaviors, individual development, and classroom practice. The predominance of these concerns has led to a rather restricted view of technology use led by enthusiasms for social-constructivist and sociocultural theories of learning. This tends to offer a very localized concept of the "social" contexts in which technology use is situated. Despite regular calls for theoretical expansion and sophistication (e.g., Hlynka & Belland, 1991; Livingstone, 2012), the educational technology research field ventures rarely from these concerns. Indeed, it could be argued that even as it approaches middle-aged respectability, educational technology is an area of academic study that remains stuck stubbornly in its ways—dominated, at best, by an optimistic desire to understand how to make an immediate difference in classrooms and, at worst, in thrall to technicist concepts of "effectiveness," "best practice," and "what works."

Of course, despite the preponderance of this sort of work, it is worth noting that other traditions of research in this field draw on cultural studies, media and communication studies, and sociology of digital culture (e.g., Facer et al., 2003; Valentine & Holloway, 2003; Jenkins, 2006). These alternative analyses are often concerned with understanding and documenting the lived digital cultures and experiences of young people and are frequently used to shine a light on the failures of mainstream education, in particular in its capacity to recognize and engage with diverse learning cultures. This research should be credited, in particular, for making visible the creativity and talents of marginalized communities in ways that disrupt the confident assertions of the school or university as monopoly authority on educational value or agent for social justice. However, such research, where it operates without a broader political and sociological account of education, risks colonization as it is recontextualized in practice. For example, celebratory ideas of young people as "digital natives" can be used to obscure the economic and social differences in young people's lives and have been recruited as justification for political projects from individualized learning to the marketization of education systems. At its worst, much of the political, cultural, and economic critique implicit in this research is lost in favor of simplified calls to appropriate digital cultural tools to engage recalcitrant youth in unchanged and unchallenged educational goals. The work becomes, too often, a set of "tips for teachers" that is disconnected from its rich, disruptive, and uncomfortable connections with the realities of life beyond the school walls. While there are researchers who resist such taming (e.g., Margolis, 2010; Mahiri, 2011) many of these who seek to retain an analysis of the politics of education and technology increasingly work outside, and are therefore arguably marginalized from, the debate about mainstream education.

As a consequence, educational technology can be a frustrating area of academic scholarship for the politically inclined reader to follow. On the one hand, thousands of hours and millions of dollars are directed each year toward the optimistic exploration of how technology is capable of supporting, assisting, and even enhancing the act of learning. On the other hand, as anyone involved with the day-to-day realities of contemporary education in its different guises will attest, many of the fundamental elements of learning and teaching remain largely untouched by the potential of educational technology. As such, an obvious disparity between rhetoric and reality runs throughout much of the past 25 years of educational technology scholarship. As Diana Laurillard (2008, p. 1) observed wryly, "Education is on the brink of being transformed through learning technologies; however, it has been on that brink for some decades now."

While similar tensions between rhetoric and reality can be found within many areas of applied academic study, a particularly resilient strain of cognitive dissonance appears to pervade the educational technology literature. Despite a long history of eagerly anticipated but largely unrealized technological transformation, many studies in the field continue to focus on the "what ifs" and "best case" examples of education and technology—often producing persuasive evidence of educational potential, but only on occasion acknowledging the individual and institutional "barriers" that are presumed to be restricting the realization of this potential in practice. As such, the academic study of educational technology could be accused of having worked itself into an analytic blind alley. This is research and writing that is well able to discuss how educational technologies *could* and *should* be used, but less competent and confident in discussing how and why educational technologies are *actually* being used. Moreover, this is research and writing that is ill equipped to support the building of an achievable political or institutional project to realize desirable change.

Such a focus on potential rather than practice also produces a number of surprising (and increasingly unforgivable) blind spots in the field. For example, while studies of the potential use of digital technologies in the teaching and learning process abound, there has been a near-universal silence on the actual use of technologies for the purpose of data collection, performance management, "learning analytics," and the like. Rather, it has taken researchers from outside the field to begin to interrogate the proliferation of data, audit, and performance management practices in which the impact of technology has arguably been spectacularly pervasive. As authors such as Jenny Ozga (2009) have noted, we are now witnessing educational systems that are based around the creation and use of data that support the organization of political relations through communication and information, with digital technologies supporting well-established systems of self-evaluation, development planning, and performance. The intentions of this data turn are deliberate—not least, the intentional move toward what Ozga (2009) terms "governing education through data" and the shift from central regulation to individual self-evaluation. In this sense, digital technologies are now an integral component of the new governance of educational institutions and those who work within them along neoliberal principles of decentralized and devolved forms of control. Educational technologists, however, have remained conspicuously silent on the role of technology in face of the "conservative modernization" of education and the emergence of a data-driven "audit culture" (Apple, 2010).

Crucially, then, educational technology remains a field of academic endeavor in which instrumentalist accounts dominate and in which the reader of much of the academic literature gains very little purchase on the social, economic, political, or cultural nature of educational technology. This matters first, because we are left with an inadequate picture of what educational technology means in and for education today. More importantly, it matters because it leaves us without the detailed and ultimately more useful accounts of educational technology that are needed to really understand and address the implications of digital technology in education for issues of social justice. Against this background, there is clearly scope to reassess and perhaps broaden the ways in which the academic study of educational technology is understood.

The question of what forms this reassessment and retuning might take is addressed throughout this book. While a number of themes are developed during the course of the book's 11 chapters, it is fairly obvious at even this early stage of our discussion that such reassessment involves moving beyond asking questions of how digital technologies "could" or "should" be used in educational settings, or speculating on the "potential" of technology to change learning. Instead, we need to take a deliberately critical approach that approaches the topic of education and technology in *relational* terms. As Michael Apple (2010) reminds us, the relational approach involves producing accounts that situate educational technology within the analysis of unequal relations of power elsewhere in society today, within the lived realities of dominance and subordination that are currently ongoing, and within the conflicts that are generated by these relations. Thus, instead of being distracted by our own (often privileged) personal experiences of digital technology, this book starts from the premise that we need to work instead toward understanding and acting on educational technology in terms of its complicated and often unjust connections to the larger society. In short, we need to develop a more politically aware and sociologically grounded narrative of change. This, then, will be the approach that shall be pursued throughout all of the proceeding chapters.

Education and Technology: The Need for a Political Perspective

This book starts from the contention that the design and use of digital technology in education is a profoundly *political* concern—inevitably raising questions of how new educational practices are being negotiated through the introduction and use of new technologies, and who benefits from such new settlements. Indeed, the use of digital technology in education introduces a complex mix of public and private actors into the education arena. These include the designers and developers of new tools, the multinational corporations that they often work for, new networks of consultants and advisors, along with new generations of ostensibly "digitally active" young people, the businesses and practices of digital youth cultures, and the families and communities within which young people lead their lives.

In this context, educational technology needs to be understood as an intense site of negotiation and struggle between these different actors. These are struggles that take place across a number of fronts—from the allocation of resources to the design of curriculum, from the maximizing of profit and political gain to attempts to mitigate patterns of exclusion. Put bluntly, as technology-based education and "e-learning" continue to grow in societal significance, it follows that the use of digital technology in education needs to be understood in distinctly political terms of societal conflict and struggle over the distribution of power. For instance, this includes acknowledging the clear linkages between educational technology use and "macro" elements of the social structure of society such as global economics, labor markets, and political and cultural institutions. Similarly, at the "micro" level of the individual, the act of technology-based learning also needs to be understood as being entwined with the "micropolitics" of social life. As such, many of the questions that surround education and digital technology are familiar from longstanding debates around education and society—in other words, questions of what education *is*, and questions of what education should *be*.

This book therefore seeks to look beyond apolitical portrayals of educational technology. Instead, it focuses on the areas of tension, contradiction, and conflict that underlie the discourses, the practices, and the technologies that constitute any instance of digital technology use in education. The forthcoming discussions therefore encompass—but are in no way restricted to—a set of interrelated issues and concerns, that is,

- the political economy of educational technology;
- the role of critical theory in the discussion of education and technology;
- educational technology, inequalities, and social justice;

- educational technology, neoliberal ideology, and emancipatory discourses;
- education technology, material resources, and resilience;
- configuring the "ideal user" of educational technologies;
- democracy, surveillance, resistance, and voice; and
- questions of who benefits from technology use in educational settings.

In exploring and examining these issues, all the chapters in this book seek to develop a set of related arguments on the theme of supporting what can be termed the *critical* study of educational technology. The role of the remainder of this introductory chapter is to provide a justification for what is to follow. As such, the chapter will now go on to outline the terms of reference for a critical approach, justifying the need for academic work that focuses on the social conflicts and politics of educational technology use at individual, institutional, and societal levels of analysis. In this sense, we seek to position the book alongside the burgeoning tradition in education scholarship for critical and democratically minded analyses of education. As Gert Biesta and others have argued, making sense of contemporary education entails focusing on a range of issues "beyond learning"—not least the political and democratic dimensions of education that are often overlooked in the relentless asking of "questions about the efficiency and effectiveness of the educational process" (Biesta, 2006, p. 22). This need to force the focus of educational technology away from the dominant discourses of "learning" and the attendant learning sciences is a fundamental task, and one that Norm Friesen and Ben Williamson undertake in the proceeding section of this book. Only by giving greater credence to critical questions of politics and democracy, it is reasoned, can academic writers and researchers then go on to develop meaningful proposals for changing educational technology provision and practice.

Steps toward the Critical Study of Educational Technology

It is important to note at this point that adopting a politically aware "critical" approach toward educational technology does not necessarily entail a dogmatic adherence to any particular theoretical stance, school-of-thought or "-ism." Rather the critical perspective is rooted in a broader recognition of technology and education as a set of profoundly political processes and practices that are usefully described in terms of issues of power, control, conflict, and resistance. As such, much of the underlying impetus for a critical approach toward educational technology stems from a desire to foster and support issues of empowerment, equality, social justice, and participatory democracy (see Gunter, 2009). These ambitions are perhaps best summarized by Amin and Thrift (2005, p. 221) in their four-point agenda for critical scholarship as follows:

> First, a powerful sense of engagement with politics and the political. Second, and following on, a consistent belief that there must be better ways of doing things than are currently found in the world. Third, a necessary orientation to a critique of power and exploitation that both blight people's current lives and stop better ways of doing things from coming into existence. Fourth, a constant and unremitting critical reflexivity towards our own practices, no one is allowed to claim that they have the one and only answer or the one and only privileged vantage point. Indeed, to make such a claim is to become a part of the problem.

As Amin and Thrift's brief manifesto suggests, a critical approach involves asking a number of questions about education and technology that seek to draw attention specifically to a critique of power and exploitation in this arena. Such a critique is not, importantly, conducted in a self-indulgent fashion for its own sake or as a detached academic exercise, but is fundamentally oriented toward opening up a better understanding of the barriers, opportunities, and resources for "a better way of doing things than are currently found in the world." To that end, we would propose the following approaches toward a critical study of educational technology.

Moving beyond a "Means-End" Way of Thinking

First and foremost, the critical study of technology and education is underpinned by a rejection of any commonsense understanding of the imperatives and potentials of educational technology or, in fact, education in general. As Boody (2001, p. 7) points out, many of the discussions of the benefits of digital technology in education take the form of "means-end thinking"—that is, thinking that starts from a given end and then strives to find the means of accomplishment. Consider, for example, the way in which the introduction of digital technologies into education has been seamlessly incorporated into narratives of educational modernization and, in more economically constrained times, to discourses of efficiency. Consider how research and policy problems have been identified as technical challenges of "harnessing" technology to existing educational goals, a formulation that has often led to a search for technologies that can be easily turned to more efficient learning "delivery." At other times, this has led to a search for tools that are, in turn, able to tame young people to achieve better compliance to existing educational imperatives. This sort of means-end thinking brings with it the elusive search for research evidence that proves the "added value" or the demonstrable cost-benefit analysis of digital technologies as compared with textbooks or teachers.

Such means-end thinking, however, fails to consider fully the nature and value of the educational end goal to which the technology is to be harnessed, and often overlooks the by-products or unintended consequences of its implementation or the connections between this given end and other important ends. Take the search for tools to enhance learner engagement, for example. A research program organized around this priority, which simply seeks to develop more entertaining or efficient ways for students to acquire predetermined knowledge, risks overlooking the multiple reasons that might cause and provoke disengagement in the first place. Resisting means-end thinking, instead, would require a critical reflection upon the assumptions underpinning an analysis of disengagement, a reflection upon alternative potential causes (such as curriculum design, staff-student relationships, and/or learning cultures) and only then begin to explore the role of digital technologies in playing a role in addressing these questions. In other words, a critical study of educational technology necessarily begins with a critical reflection upon the definition of the educational "problem" at hand.

Opening up the Black Boxes of Educational Technology

The second key feature of a critical approach is its commitment to disrupting the deterministic assumption that technologies possess inherent qualities and are therefore capable of having particular predetermined and predictable "impacts" or "effects" on learners, teachers, and wider society. There are two aspects to this commitment: first, it entails a critique of the logic of inevitable sociotechnical change, and second, it entails a surfacing of the politics, contradictions, and negotiations realized in technologies and technological practices.

Of course, we are not the first (or last) authors to decry the persistence of the technological determinist mind-set in educational technology studies (see Bromley, 1997; Oliver, 2011). The technologically determinist perspective that "social progress is driven by technological innovation, which in turn follows an 'inevitable' course" (Smith, 1994, p. 38) has a long lineage in popular and academic understandings of the effects of technology on society. It produces an understanding that individuals and institutions should simply "adapt to" technological change and make best use of the technologies that they are presented with. This commonplace logic—evident in much thinking about educational technology—is illustrated in Clay Shirky's (2008, p. 307) observation that

> our control over [digital] tools is much more like steering a kayak. We are being pushed rapidly down a route largely determined by the technological environment. We have a small degree of control over the spread of these tools, but that control does not extend to being able to reverse, or even radically alter, the direction we're moving in.

Such determinist thinking underpins a range of different positions in educational technology. On the one hand, for some, it underpins popular claims that video games *cause* violent behavior or that social media use *leads* to antisocial behavior. These claims lead to calls to ban all computer games and prevent young people from accessing the internet. On the other hand, such determinist thinking can also be found in popular claims that new technologies are changing the world and that, in turn, educational institutions have no choice but to adopt digital technologies, designed elsewhere, for other purposes, for fear of being "left behind."

Any history of the development and appropriation of digital (or indeed any) technologies, however, makes such a determinist position untenable. Whether thinking about the development of electricity or the bicycle, the record player or the computer, there is now over 30 years of research making clear that the course of technological development is neither inevitable nor predictable and that the complex processes of using and appropriating emerging tools will lead to unexpected and divergent future trajectories (Woolgar, 2002). It is crucial to acknowledge from the outset, therefore, that there is neither an inevitable "technological future" to which schools need to adapt, nor a set of universal technological impacts from which young people need to be protected (Facer, 2011).

As such, the critical study of educational technology starts from the premise that "devices and machines are not things 'out there' that invade life" (Nye, 2007, p. ix). More emphasis is placed on understanding the development and implementation of technological innovations as set within specific social and economic contexts, instead of new technologies somehow having inevitable internal logics of development regardless of circumstance (see Williams, 1974). Following this line of argument, the critical study of educational technology accepts that there can be no predetermined outcomes to the development and implementation of educational technologies. Instead, any technological artifact is seen as being subjected continually to a series of complex interactions and negotiations with the social, economic, political, and cultural contexts that it emerges into. Before a technological artifact is used (or not) by a learner, for example, the said technology will have been party to a complex of vested "other" interests above and beyond the actions of its initial designers and producers. These other interests range from marketers and journalists to (quasi) government agencies, teacher unions, and consumer interest groups—all having a significant but often subtle bearing on the shaping of educational technology, and all therefore meriting sustained scrutiny and questioning. Understanding technology as being "socially shaped" therefore allows for analyses that "open up the black box of technology" (Bijker, Hughes & Pinch, 1987), which consider the organizational, political, economic, and cultural factors that pattern the design, development, production, marketing, implementation, *and* "end use" of a technological artifact (see Selwyn, 2012).

As Langdon Winner (1986) puts it, at the heart of this approach is the recognition of the politics of educational artifacts. From this perspective, tools cannot be understood as "neutral." Indeed, as Bijker and Law (1992, p. 3) assert, most people would agree that "the idea of a 'pure' technology is nonsense." Instead, educational technologies are best understood as socially constructed, shaped, and negotiated by a range of actors and interests both in their construction and procurement and in their realization and use in practice. Technological artifacts can be understood as political because they are, in material form and in practice, the outcome of competing agendas. Put simply, technological artifacts have an "internal politics." Yet, it is also important to acknowledge that these technologies also have "external politics"—that is, "they are designed to do things, to allow some behaviors and uses, and to prevent others" (Matthewman, 2011, p. 5). All technologies are designed, created, and implemented to create a type of order or settle a dispute. They are often designed to fit with a particular type of political arrangement. In a context of radical economic, political, and educational inequality, technology is usually arranged around certain values, power is centralized, hierarchies are embedded, allocation is uneven, and there are structural constraints between social classes. All of these issues constitute what Smits (2001) described as "a shadow constitution" of hidden political power that pervades any technological form.

Crucially, as we progress through the second decade of the twenty-first century it is important to recognize that the "political technologies" of today are not the obvious technologies of class struggle throughout the twentieth century—what Virilio (2003) calls the "big machines" of the industrial era, for example, the mills, the factories, the military machinery. Instead, they are the small devices of the postindustrial "technosphere"—that is, the digital devices and gadgets that now permeate our everyday lives. The politics of these technologies are often difficult to identify and scrutinize—as Steve Matthewman (2011, p. 72) argues, "Because these beliefs are embedded in the very fabric of our technologies we often fail to interrogate them seriously. Yet question them we must." A critical study of educational technology, then, is a study that treats the "tools" themselves not as black boxes with inevitable social and educational impacts, but as contingent, provisional settlements between social actors in which the wider political conditions of the period are realized and may be resisted. As Wiebe Bijker (1995) reasons, only by exploring and exposing the social roots of such technology can academics hope to make the technology amenable to democratic interpretation and intervention.

Asking "State-of-the-Actual" Questions

As implied at the beginning of this chapter, the academic study of educational technology is often drawn inexorably toward a forward-looking, "leading-edge" perspective. Digital technologies are often used as a proxy signifier for "the future" and for education's capacity to adapt to and prepare for the future. As such, many educational technology commentators, writers, and researchers tend to show most interest in what could be termed "state-of-the-art" issues—addressing questions of what might happen once the latest technologies, sufficient economic resources, and expert practitioners are placed into educational settings.

As David Buckingham (2007) and others have observed, the educational technology literature therefore abounds with in-depth investigations of "model" education institutions and classrooms with enthusiastic tutors and well-resourced students basking in the glow of the "Hawthorne effect" of the attention of researchers. Such research tends to cast the educational technology researcher as futurist, providing insights into worlds to come, or as designer and creator of new realities. Yet the practical significance of such avowedly state-of-the-art perspectives on new technology and education are often limited. This work offers little useful purchase on contemporary problems or insight into how present arrangements may be improved or ameliorated—usually resulting in little more than a tentative "proof of concept" that is woefully decontextualized and unrealistic.

In contrast, the critical study of educational technology retains a firm "commitment to the here and now." It reframes the problem of the future by arguing, with Miller (2011, p. 1), that "the challenge is not that we must find ways to 'know' the future, rather we need to find ways to live and act with not-knowing the future." Such an approach, as Brian Massumi observes, locates possibility and potential not in the attainment of some future utopia, but in the messy realities of the present.

In every situation there are any number of levels of organization and tendencies in play, in cooperation with each other or at cross-purposes. The way all the elements interrelate is so complex that it isn't necessarily comprehensible in one go. There's always a sort of vagueness surrounding the situation, an uncertainty about where you might be able to go and what you might be able to do once you exit that particular context. This uncertainty can actually be empowering—once you realize that it gives you a margin of maneuverability and you focus on that, rather than on projecting success or failure. It gives you the feeling that there is always an opening to experiment, to try and see. This brings a sense of potential to the situation. The present's "boundary condition," to borrow a phrase from science, is never a closed door. It is an open threshold—a threshold of potential. (Massumi, cited in Zournazi 2002, p.211)

This temporal orientation for critical study of educational technology therefore has a number of methodological implications. In particular, it implies a need for accounts of digital technology that concentrate on developing "thick" descriptions of the present uses of technologies in situ rather than speculative predictions and forecasts of the near future. These can be seen as state-of-the-actual questions—that is, questions concerning "just what is going on" when a digital technology meets an educational setting and what institutions, histories, agents, tools, and concepts are and have been in play in this process. This approach therefore draws on traditions of detailed qualitative research and invokes theoretical perspectives that encourage attention to the rich sociomaterial practices in play and to the negotiations, conflicts, and struggles between social actors in those settings (Fenwick & Edwards, 2010). Such a perspective resists the often-abstracted approach to studying educational technology in which, as Charles Crook (2008) observes, the "spontaneous appropriation" of digital technologies by students, teachers, and others is often assumed. Instead, it encourages attention to the microlevel interactions between educators and "learners" and to the ways the local contexts frame learning processes and practices. It recognizes that technology-based learning enters existing histories and cannot be "detached from the spatial condition of common locality" (Thompson, 1995, p. 32). In this manner, the critical approach seeks to foreground attention to the way in which educational and emancipatory possibilities are realized and closed off, not in a future world, but through the choices and practices of educators, students, policy makers, and commercial companies in the messy realities of educational institutions today.

Asking Who Benefits?

The attention of the critical researcher toward the messy realities of actual practice is not oriented simply toward description or toward educational settings disconnected from wider society. Instead, it is focused on the analysis of how this reality is implicated in reproducing or unsettling wider patterns of educational, social, and economic inequality. In other words, the critical researcher is concerned with understanding "who benefits" from the introduction of educational technology. This concern therefore directs our attention toward a number of broader issues that articulate with the practices and policies of educational institutions; inter alia, it encourages attention to the way in which scarce resources are allocated in the education arena, to the way in which existing institutions of power and wealth are connected to educational practices, and to the functioning of marketplaces and the role of commerce and commercial actors in the educational technology field. It also focuses attention on the relationship between the "education industry" and the institutions of advanced capitalism, in particular, the function of education institutions in shaping and providing workforces for the labor market.

Such interests can be understood as constituting a political economy of the educational technology field. As Vincent Mosco (2009, p. 4) puts it, "The political economist asks, how are power and wealth related and how are these in turn connected to cultural and social life?" Of course, it is important not be seduced into a total state of "economism" where issues of economics and economy are allowed to overshadow all other issues. As Bernard Stiegler (2009, p. 7) reminds us, for the political economist "the question is as political as it is economic." Therefore, the best political economy accounts aim to unpack what Stiegler terms "the totality of social relations" between economic, political, social, and cultural areas of life.

One of the benefits of taking this approach is to make explicit issues of power within an area of society such as educational technology. Political-economy analysts will often concern themselves with developing accounts of emerging and established hierarchies of power and providing explanations for their legitimation. In terms of education and technology, then, the political-economy approach encourages an interest in the ways in which structures and processes of power are embedded within digital technology products and practices, as well as how the lives of individuals are then mediated by educational technologies. Important here are questions of domination, subordination, and how the use of digital technologies in education contribute to the perpetuation of existing—and often deeply rooted—inequalities. As Robin Mansell (2004, p. 98) reasons, "If resources are scarce, and if power is unequally distributed in society, then the key issue is how these scarce resources are allocated and controlled, and with what consequences for human action."

One of the key insights to be gained from political economy approach is the linkage between educational technology and the interests of capital and capitalism. Indeed, political economy commentators are traditionally interested in questions of production, consumption, work, labor, industry, marketing, and commerce (Stiegler, 2009). At one level, then, the political economy approach focuses attention on the workings of the education industry—raising questions of how the "business" of education operates and the ways that particular forms of innovation (such as digital technology) are "recruited, put to work and traded upon" (Apple, 2010, p. 30). The critical researcher, then, would use the political economy approach to raise concerns over the commercialization of technology-based education and the state-approved (and even state-sponsored) liberalization of educational technology markets to widespread global competition. It would also encourage questions about the associated internationalization of authority as national educational authorities cede control and power over educational technology arrangements to regional alliances and authorities. In particular, then, the critical study of educational technology raises questions of how digital technology is implicated in educational circuits of production, distribution, and consumption. Through all of these analyses, the critical researcher draws on political economy analyses to understand who benefits from these changes, to make visible how, where, and whether the existing relations of inequality are being reproduced and unsettled through these disruptions to educational institutions and practices.

Identifying Opportunities for More Equitable Futures

As much of the previous discussion has implied, the impetus for taking a critical approach toward the study of educational technology is rooted in the high-minded but well-intentioned aim of making education *fairer* as well as merely better or more "effective." Critical analyses therefore seek to address the fact that the use of digital technology in educational settings is often not a wholly inclusive, dialogic, or equitable process in which all actors have equal power in participating, and where all actors can determine what educational technology is or how it is used. The critical take on educational technology is therefore often driven by a desire to redress the imbalances of power that reside within most educational uses of technology. Thus, instead of indulging in what C. Wright Mills (1959) derided as "abstracted empiricism," a core concern of the critical approach is to identify, highlight, and overcome the many contradictions and conflicts that surround the use of technology in educational settings. The critical study of educational technology is therefore pursued with an overarching intention of developing culturally plausible suggestions as to how current inequalities and hegemonies may be countered, and how digital technology use in educational settings may be reshaped along fairer and more equitable lines.

This suggests a tradition of educational technology scholarship that builds upon Ann Oakley's (2000) notion of social science research that is democratic, interventionist, and emancipatory. In this spirit, the critical study of educational technology can be used to work with educators, students, and communities to identify spaces where opportunities exist to resist, disrupt, and alter the technology-based reproduction of the "power differential that runs through capitalist society" (Kirkpatrick, 2004, p. 10). Rather than locating its hope in an abstract future to be realized once all necessary economic and technological conditions are met, critical scholarship in this area instead draws attention to the creativity of the present as a resource for optimism.

Indeed, it is important to remember throughout the course of the book that these analyses are not meant to be defeatist in their outcomes. Ultimately, we have approached the writing and reading of the book in the hope of it being a constructive rather than destructive exercise. Throughout this book, we must remain mindful of the need to not merely decry the unsatisfactory state of the present, but also to consider opportunities and spaces for future critical action as well as critical scholarship. As such all the chapters in the book have been written in the spirit of offering an analysis that is able to point toward contradictions, controversies, *and* the spaces of possible action. Such an orientation is intended not to establish a new orthodoxy of educational technology (as if that would ever be possible), but rather to open up new spaces for debate about the history, practice, and opportunities for action in this arena. It is intended to unsettle the dominance of the means-end discourse of educational technology in ways that begin to allow a much wider range of social actors to claim the right to speak and contribute to the shaping of the technologies that we use in education. It is intended as the beginning, rather than the end of a conversation.

Conclusion

This chapter has attempted to make the case for a context-rich analysis of educational technology. It has argued for a research orientation that pays attention to

- the discourses and purposes of education within which educational technologies are framed, and the definitions of the problems to which they are invoked as solutions;

- the contested and unstable political processes by which "technologies" are designed, introduced, and appropriated within education;
- the messy realities of actual educational technology practice in the institutions of today;
- the political economy of the educational technology field and its implications for social justice; and
- that seeks to open up new spaces for debate with a wide range of social actors about the practices and policies of educational technology.

The chapter has shown how a critical approach allows a number of "big questions" to be asked about technology and education—not least how individual learning technologies fit into wider sociotechnical systems and networks, as well as what connections and linkages exist between educational technology and macrolevel concerns of national and global economics. In contrast to these grand concerns, the chapter has also shown how a critical approach offers a direct "way in" to unpacking the social processes that underpin the microlevel politics of digital technology use in educational settings. From both these perspectives, the principal advantage of a critical approach should be seen as the ability to develop a more socially grounded understanding of the messy realities of educational technology "as it happens." In approaching education and technology as a site of intense social conflict, a critical approach allows researchers and writers to address questions of how digital technologies (re)produce social relations, in whose interests they serve, and identify sites for resisting and unsettling such relations (see Apple, 2004).

In extolling the virtues of a critical stance on education and technology this book's intention is not to indulge in academic one-upmanship or convey an arrogant belief that one particular intellectual approach is more privileged and correct than any other. Indeed, none of the chapters in this book set out to contend that their critical approach is somehow superior to existing modes of inquiry and analysis. Rather, it should be concluded that all of the chapters in this collection offer an important additional dimension to the study of educational technology—providing an often challenging but ultimately complementary perspective to the learning-centered studies that have dominated the field over the past 30 years or so. As such, we hope that this book is received by the educational technology community in the spirit that it was intended—as a means of broadening the scope of the conversations and debates that surround the field as it enters a significant period of mainstream prominence.

Moreover, despite the grand scale of all its aims and issues, we hope that the book nevertheless retains a clarity and usefulness in its overall analysis.

Although one of the main aims of writing this book was to problematize the universalizing nature of the discourses that have come to surround technology use in education, it is important not to be overwhelmed by the scope and diversity of the issues that are under discussion. One dominant theme to emerge by the end of this book is that while the educational technology is a topic that is beset by conflict and contradiction, there *is* still a clear story that can be told about education and technology. As we shall see, this is a story that is as much as about conflict as it is as about change; and that is as much about ideology as it is about innovation.

PART I

Recognizing the Politics of "Learning" and Technology

CHAPTER 2

Educational Technology and the "New Language of Learning": Lineage and Limitations

Norm Friesen

Introduction

Even the most rudimentary definitions of the term "technology" indicate that its meaning extends far beyond artifacts and devices to include processes, methods, means, and applied knowledge. It is therefore surprising how rarely instructional theories, methods, and applications—for example, learning theories, learning designs, or learning environments—are considered specifically as technologies in the relevant literature. This chapter focuses on the instrumental nature of the concepts of "learning" and "learning theory" as used in the field of education and technology—which Gert Biesta (2006) and others have characterized as being manifest in a "new language of learning." This refers to a vocabulary or discourse that, for example, characterizes "teaching [as the] 'facilitation of learning' [and], education [as the] 'provision of learning opportunities'" (Haugsbakk & Nordkvelle, 2007, p. 2). The chapter argues that this vocabulary represents a particular technologization or instrumentalization of education, a process that makes educational practices and priorities appear germane to, or even incomplete without, technological rationalization and reshaping. This chapter traces how this vocabulary casts learning as a natural and universal process, and quite consistently accompanies the promotion of a range of technological artifacts in education. Running from the introduction of "teaching machines" through to current visions of school reform, this theoretical lexicon will be shown to efface its cultural and ideological contingency through a quasiscientific "neutrality" and a biologically based universality, and to limit the possibility for discourse and practice within the field of education and technology.

The New Language of Learning

It has become relatively uncontroversial to depict teaching and education as a means through which naturally occurring, biologically based processes of learning are directed and facilitated in order to achieve predetermined outcomes. Gert Biesta, who has described this situation in terms of the new language of learning, attributes its recent emergence to a multiplicity of factors:

> It is important to see that the new language of learning is *not* the outcome of a particular process or the expression of a single underlying agenda...There are at least four trends that, in one way or another, have contributed to the rise of the new language of learning. (2006, p. 17; emphasis in original)

Biesta's four trends are economic ("the erosion of the welfare state"), demographic ("individualistic and individualized adult learning"), broadly philosophical (a questioning of the modernist project of education), and disciplinary ("new theories of learning"). While I do not question the importance of any one of the factors listed by Biesta, it is the fourth trend that appears to have the longest direct lineage. Certainly within educational technology as a field, this trend has exercised the strongest influence on discourse as well as priorities and practice.

Biesta describes new theories of learning in terms of "the emergence of constructivist and sociocultural theories" that see learning principally as an activity on the part of the student—and that reduce the role of the teacher to one of support and facilitation:

> Such theories have challenged the idea that learning is the passive intake of information and have instead argued that knowledge and understanding are actively constructed by the learner, often in cooperation with other learners. This has shifted the attention away from the activities of the teachers to the activities of the students. As a result, learning has become much more central in the understanding of the process of education. (2006, p. 17)

While these constructivist and sociocultural approaches emerged around the 1990s, the theoretical lineage of the "language of learning" extends much further back than this. This chapter traces this lineage back to the very beginnings of educational psychology in the early twentieth century, and shows how a language of learning—in its different manifestations—has shaped and limited discourse in education and educational technology ever since. For example, between 1900 and the 1920s learning came to be defined as a kind of discrete and naturally occurring process, one that is to be optimized primarily through practice and repetition. Learning as a process was therefore grounded firmly in a mechanistic, causal frame of reference, and this grounding was preserved in the context of the cognitive revolution, which recast learning as computational processing. Later—in conjunction with Biesta's sociocultural and constructivist influences—conceptions of learning were modified to appear as computationally tractable, representational "knowledge building." After tracing these developments in the language of learning, this chapter considers the broader implications of the language of learning in educational contexts, arguing that like material technologies, it is a social construction rather than a simply "natural" biological fact, and as such, it has the power to limit and distort understandings of what is possible and necessary in education.

Thorndike: The "Animal Method of Learning"

Before the end of the nineteenth century, learning generally referred to a doctrine of teaching, or to "the action of receiving instruction or acquiring knowledge," typically associated with teaching, life experience, or inherited tradition (OED, n.d.). The vicissitudes of the last of these—inherited tradition—were the target of Bacon's *Advancement of Learning* (1605); and many of the other traditional meanings are captured in scientist Richard Owen's 1862 remark that "there's nothing so good for learning, as teaching." However, in the first decade of the twentieth century, Edward L. Thorndike, "perhaps the greatest learning theorist of all time" (Hergenhahn & Olson, 2001, p. 51), helped introduce a new meaning of the term. Thorndike defined learning as a naturally occurring process common to both humans and animals. In a book titled *The Human Nature Club: An Introduction to the Study of Mental Life*, Thorndike mentions for the first time a "method of learning" that humans share with "many animals besides man":

> The gradual increase in success means a gradual strengthening of one set of nerve-connections, and a gradual weakening of others. This method of learning...is the most fundamental method of learning...[and] may

be called the method of trial and error, or of trial and success, or (from its importance in animal life), *the animal method of learning*. (Thorndike, 1901, p. 38; emphasis added)

The idea that humans, like other mammals, learn through a common, underlying mechanism involving simple rules (trial and error) laid the foundation for what was later to become Thorndike's connectionist theory of stimulus and response. As such, learning provided the foundation for a systematic program of psychological and educational research that Thorndike pursued throughout his highly prolific and influential career. In texts discussing "processes of learning," "general laws of learning," or the acceleration of the "rate of learning," Thorndike was the first to speak extensively and comprehensively of learning as a set of behaviors to be systematized and manipulated with predictive precision through experimentation.

Thorndike experimented with cats, dogs, and other caged animals, timing their repeated attempts to access food by pressing on a lever or pulling on rope. From this, he established the first of his "laws of acquired behavior or learning" (1911, p. 244)—the law of effect. This he described as "[responses] accompanied or closely followed by satisfaction to [an] animal will, other things being equal, be more firmly connected with the situation, so that, when it recurs, they will be more likely to recur." Writing in *Education: A First Book* (a textbook for teachers still in print today), Thorndike presents a series of laws of learning, which he asserts govern the learning of any specific subject that education might seek to support. Based on these laws, Thorndike asserts, "any problem of education may be put in the form, 'Given a certain desired change in a man, what situation shall we create to produce it, either directly or by the response which it provokes from him?'" (1912, pp. 55–56). More complex manifestations of learning and development, such as character, ability, or knowledgability are similarly reducible to situation and recurrent response, "what we call intellect, character and skill," as Thorndike explains, is "in the case of any man, the sum of the man's tendencies to respond to situations and elements of situations" (1912, p. 102).

As a side note, if the displacement or questioning of the human as a category is constitutive of the "posthuman," then Thorndike and learning theorists after him articulated a convincingly posthumanist position many decades before the term was coined. The posthuman compatibility between human and animal, and (as will be shown below), human and machine is repeatedly underscored in Thorndike's "animal learning" model and in the learning theories that followed. But unlike the "information-rich" environments visualized by cognitive learning theorists, the paradigmatic site of learning for Thorndike is a naked creature in trial and error reflexive response to its immediate environment. The organism (a human or other type of mammal), like the context in which it is embedded, is stripped of any interpersonal, historical, cultural, or political significance. Also, when all learning is reduced to these terms, the function of teaching appears as little more than a relatively simple set of operations that can be undertaken just as easily by a human as by a machine. Indeed, Thorndike envisions such a machine as early as 1912:

> If, by a miracle of mechanical ingenuity, a book could be so arranged that only to him who had done what was directed on page one would page two become visible, and so on, much that now requires personal instruction could be managed by print. (Thorndike, 1912, p. 165)

These possibilities were to be realized some 50 years later in the radical behaviorism of B. F. Skinner. Skinner's work was based on Thorndike's reduction of learning to behaviors acquired through stimulus and response. Skinner, however, focused on changes of behavior produced through programmatic "reinforcement." This reinforcement can occur through consequences or changes that are introduced in the immediate environment in a manner systematically "contingent" upon behavior, leading to the Skinner's notion of "contingencies of reinforcement." Combined with observable changes in behavior, contingent reinforcement provides Skinner with what is required to describe the function of instruction or teaching as a technology or technique to be executed by either human or machine. "Teaching as a technology," he explains, operates through the arrangement of "contingencies of reinforcement under which behavior changes" (1968, p. 9).

Having reduced education to these minimalist, causal terms, Skinner was able to conceptualize a type of technologized teaching in the form of a mass produced machine that could fit on students' desks, and that could operate independently according to each student's abilities. These devices, which came to be known as teaching machines, provided stimuli in the form of small units of information and immediate, consistent, and programmed feedback or reinforcement. Teaching complex subjects—or "program[ming] complex forms of behavior" according to Skinner—was simply a question of the "effective sequencing" of mechanical inputs and feedback. "We have every reason to expect," Skinner concludes, "that the most effective control of human learning will require instrumental aid. The simple fact is that, as a mere reinforcing mechanism, the teacher is out of date" (Skinner, 1968, p. 22).

Although it is only infrequently mentioned today, the "programed instruction movement" or "teaching machine revolution" (Saettler, 2004, p. 294) that accompanied Skinner's bold pronouncements set an important precedent for learning theories that followed Thorndike's connectionism. By mobilizing a limited set of highly simplified elementary components (e.g., stimulus, response, and reinforcement), learning becomes described in terms that are universal, causally and biologically oriented. In this way, the differences between the various forms and experiences associated with learning—and their historical, cultural, and political contexts—are effectively eliminated. In their place, a technological device or "solution" is advocated, which is able to satisfy the requirements for learning thus defined. In the particular example of Skinner's language of learning and his teaching machine, technology as a manufactured artifact and as a theoretical or discursive manifestation—illustrated through references to reinforcement "mechanisms" and "technolog[ies] of teaching"—can be said to be so closely aligned as to be inseparable.

Cognitive Psychology: Infantalization of Education?

Cognitivism took over as the dominant paradigm for learning theory in the context of the "cognitive revolution" of the 1970s. In place of environmental stimuli and neural responses, cognitive learning theory sees the mind as an "information processing system," a premise still in evidence in the writing of many learning theorists today. Like a computer, the working of the mind can be understood in terms of informational inputs and outputs, and internal computational "states":

> Cognitive learning is equated with discrete changes between states of knowledge rather than changes in the probability of response. In cognitive learning, issues of how information is received, organized, stored, and retrieved by the mind is important…The most dominant of the cognitive learning theories is based on an information-processing approach. (Uden & Beaumont, 2006, p. 6)

One event that precipitated the rapid spread of cognitivism in psychology was a review of Skinner's book *Verbal Behavior* published by linguist Noam Chomsky in 1967. In this article, Chomsky argued with "devastating clarity" that human language is grammatically much too complex to be acquired through processes of stimulus and response (Chomsky, 2006, p. 244), and that it is with the complexity of syntax rules that any study of language acquisition must begin:

> It appears that we recognize a new item as a sentence not because it matches some familiar item in any simple way, but because it is *generated*

by the *grammar* that each individual has somehow and in some form *internalized*. And we understand a new sentence, in part, because we are somehow capable of determining the process by which this sentence is derived in this grammar. (Chomsky, 1967, p. 143; emphases added)

We are able to use language, in other words, not because of familiar patterns of stimulus and response, but because we possess an "internalized" ability to encode and decode data according to complex syntactical rules. Chomsky later developed these intuitions into a full-fledged theory of universal, generative grammar, and into the notion of an innate, mental "language acquisition device." In these terms, generative grammar is a complex set of "deep structures" or rules underlying all human language; and the language acquisition device is a faculty (a "language organ") that Chomsky believed allowed infants to internalize linguistic rules derived from a general grammar. Infants could do this most effectively, Chomsky believed, if the information in which they were placed was data or information rich—ideally if the infant's caregivers were themselves speaking. Through this rationalist, "computationalist" account of language acquisition and comprehension, Chomsky offered a compelling paradigm case—some even say a "particularly powerful ideology" (Golumbia, 2009, p. 31)—for what followed in educational psychology and learning theory.

As in behaviorism, following Chomsky, learning is constructed in cognitive theory as something that occurs naturally, is fundamentally individualized (taking place through the workings of one or more organs), and can be broadly subsumed under one operation, the processing of information or data. A paradigm case for this process, as provided by Chomsky, is natural learning of language in infancy. Citing Jean Piaget in addition to Chomsky, Christina Erneling (2010) argues that these "innatist," "rationalist," and "universalist" convictions concerning learning have led to notable results for education in general:

> Pedagogical thinking has appropriated from developmental cognitive psychology...the idea that all learning is like early infant learning, that is, all learning is grounded in biological abilities and is to a large extent innate, automatic, and unconscious. The task of the school is to mimic the conditions of this early learning situation so that learning in schools will improve. (Erneling, 2010, p. 4)

Erneling concludes pithily that this cognitivist, Chomskian theory of learning "infantilizes" education as both a set of practices and an object of study. Learning is conceived of as a natural process that occurs according to an explicitly general, universal set of rules. Whether this restricted to the "deep processing" of a universal, generative grammar, or includes the later cognitivist constructs such as mental mapping, self-regulation, or other quasicomputational processes, cognitive learning theory casts the role of education as a means of augmenting or facilitating human computational capacities. Again, the historical, cultural, and political dimensions of education are effectively collapsed, and in place of its roles, values, and institutions, the most expedient solution to be put in their place, not surprisingly, is computational learning technologies themselves. Apparently unaware of the circularity of this way of thinking, one educationalist promoted the use of computers in "the learning process" as follows: "To be effective, a tool for learning must closely parallel the learning process; and the computer, as an information processor, could hardly be better suited for this" (Kozma, 1987, p. 22).

In her discussion of cognitivism's "infantalization" of education, Erneling also shows that prominent educational technologists have leveraged the computational or rationalist models of Chomsky and Piaget to make the case for the educational efficacy of specific computer technologies. Seymour Papert and Roger Schank, for example, argued that the conditions of early, "deep" learning are mimicked most effectively in school through the use of specific computer technologies with which students can (often literally) play. Writing in his famous *Mindstorms* book, Papert (1980), for example, argues that

> the model of successful learning is the way a child learns to talk, a process that takes place without deliberate and organized teaching. I see the classroom as an artificial and inefficient learning environment... [and] I believe that the computer presence will enable us to so modify the learning environment outside the classrooms that much if not all the knowledge schools presently try to teach with such pain and expense and such limited success will be learned, as the child learns to talk, painlessly, successfully, and without organized instruction. (pp. 8–9)

Roger Schank makes a similar case in *Teaching Minds: How Cognitive Science Can Save Our Schools* (2011), in which he generalizes the conditions of infant learning as being broadly "experiential." All learning in Schank's view should ultimately take the form of "learning-by-doing." Early in his text Schank focuses on the example of learning the skills of "walking and talking," pointing out that such acts are not only "intrinsically rewarding," but also that they can be learned only by engaging in the activities themselves. Here and elsewhere, Schank envisions how such "hands-on" learning can be mobilized in the form "online, learning-by-doing, experience based, learning environment[s, with] teaching occurs on an as-needed basis" (Schank, 2000 p. 590). In such practice-based learning, the role of the teacher or instructor, in Schank's (2011, p. 7) view, can only be one of facilitation: "The only teaching that can work, then, is the kind of mentoring that helps someone execute better what they are practicing."

Cognitivism plays a foundational role in the constructivist and sociocultural psychological theories that developed in the decades following the cognitive revolution, and that are directly referenced in Biesta's original description of the new language of learning. Both constructivist and sociocultural theories arguably have their origin in the influential work of Russian psychologist Lev Vygotsky. This influence can be traced specifically to Vygotsky's proposal that linguistic signs or symbols are not simply neutral carriers of information but that they act as "instrument[s] of psychological activity in a manner analogous to the role of [tools] in labor" (Vygotsky, 1978, p. 52). Constructivist and sociocultural learning theorists have subsequently reasoned that "cognitive" tools—whether they are primitive linguistic symbols or powerful computational "symbol manipulators"—can help form and shape the fundamental character of the laborer or tool user. Despite the potentially radical nature of Vygotsky's insight,[1] his study of learning as a constructive activity involving linguistic tools has provided the opportunity to imagine the use of the computer as a particularly powerful "cognitive device" or "mindtool." Simply put, the significance of education in this particular discourse is defined not in terms of facilitating the deep processing of data or information, but rather in terms of the representation, exemplification, or modeling of that which is inherently mental.

David Jonassen, a name associated with both cognitivist (e.g., Jonassen, Hannum & Tessmer, 1989) and constructivist (e.g., Jonassen, Peck & Wilson, 1999) approaches, explains,

> Succinctly, constructivism avers that learners construct their own reality [on the basis of their] prior experiences, mental structures, and beliefs...What someone knows is grounded in perception of physical and social experiences which are comprehended by the mind. What the mind produces are mental models that represent what the knower has perceived. (Jonassen, 1994, pp. 34–35)

The knowledge thus constructed is then judged in terms of its practical *viability*. In other words, knowledge can be validated "by testing the extent to which it provides a workable, acceptable action relative to potential alternatives" (Duffy & Cunningham, 1996, p. 171). Propositions or hypotheses, as one example, can be verified or falsified in collaborative learning through reference to new, previous, or shared experience. In other educational contexts, learner (or "novice") constructions, mental models, or interpretations, can be compared with those of professionals or specialists, and brought into closer alignment with these "expert" constructions, "schemas," or "models" (e.g., Bransford, Brown & Cocking, 2000).

In this context, the educational contribution "of interactive multimedia, animation, and computer modeling technologies" can be readily defined in terms of their "representational affordances" (Jacobsen, 2004, p. 41). These affordances are of value particularly because they allow for the "dynamic qualitative representations of the mental models held by experts and novice learners" to be visualized and easily compared (Jacobsen, 2004, p. 41). This in turn allows for the gradual approximation of novice models or schemas to those of more expert constructors of knowledge. Similar analogies or arguments have been made in the case of other representational capabilities of the computer. These extend from hypertext to hypermedia (e.g., Jonassen et al., 1989) and from computer supported collaborative learning environments (e.g., Hoadley, 2005), to more interactive and kinesthetic forms of games and simulations (e.g., Gee, 2007). In each case, it is the ability of advanced computer technologies to represent knowledge—and thus make "normally hidden knowledge processes…transparent to users" (Scardamalia, 2003, p. 23)—that provides the terms for defining the value of technologies and, by extension, the purposes of education itself.

The result of this combination of Vygotsky's sociocultural theory with the rationalist universalism of the cognitive sciences is that the social and cultural specificity of Vygotsky's understanding of "tools" and signs is effectively nullified. Tools no longer have the contingency, complexity, and ambivalence of changing devices and their configurations, linguistic signs, and historically layered etymologies. Instead, social and cultural considerations in education are actually erased in favor of purified instrumentality that is bereft of cultural and ideological contingency. A tool, whether a word, pencil and chapter, or a networked computer, has educational value through its representational affordances, its ability to mirror the mental representations or schemas, which are seen as the basis for learning in constructivist learning theory. In addition, it is not surprising to learn that the representational affordances of these various tools are at their most efficient and effective in multimedia computers that became popular in the 1990s. Unlike other tools of representation, the PC technology of the 1990s afforded representations, structures, and schema that were multidimensional, multimedial, and dynamic (i.e., open to endless revision). As the term "mindtool" suggests, the constituents of the paradigmatic scenario for learning in this context is a kind of disembodied connection of mind with computationally enabled representational device. A similar nomenclature is utilized in the emerging interdisciplinary "learning sciences." Casting this "new science" as an eclectic, but ultimately positivistic attempt to both "understand and propel human learning," self-identified learning scientists confidently predict that a "mature science of learning" will soon "discover its neural underpinnings and identify the internal mechanisms that govern learning across ages and settings" (Bransford et al., 2006b, pp. 210, 212). Like other approaches to learning before it, this new science is premised on a notion of learning as biologically grounded, governed through internal mechanisms and marked by sufficient invariance to allow it to be first isolated and then exploited through predictive scientific means. Like behaviorism and the cognitivist revolution before it, this recent development in learning theory is also closely tied to new technological products and developments. However, it frames these in terms much more general than the introduction of a single device, computer technology, or set of applications. It sets as its goal nothing less than the "mammoth undertaking" of "redesigning schools so that they are based on scientific research," rather than on outmoded social conventions. The technology of the classroom, of instructional methods and techniques, and of everyday educational practices is to be subjected to a rationalized redesign. Many of the visions of such redesigned schools show a surprising alignment with changes advocated on the conservative end of the political spectrum—particularly in the United States:

> To take just one hypothetical possibility, tutoring centers like Sylvan Learning Centers [a US-based K-12 "supplemental learning service"] might begin to offer a three-hour intensive workday, structured around tutors and individualized educational software, with each student taking home his or her laptop to complete the remainder of the day at home. (Sawyer, 2006, p. 569)

Like other educational technologists and reformers before them, constructivist-minded learning scientists have described the school as a hopelessly outmoded industrial or even preindustrial institution in an age that dropped these social and economic forms. They foresee the economic "rationalization" that has marked the landscape of northern England or the northeastern United States as occurring more broadly in education, "leaving today's big high schools as empty as the shuttered steel factories of the faded industrial economy" (Sawyer, 2000, p. 569).

Although its vision of the future of education differs from those of Skinner or Schank, the recent efforts of learning science to "understand and propel human learning…across ages and settings" (Sawyer, 2000, p. 569) is part of a century's worth of natural scientific effort to identify the essentials of an innate and universal process or set of processes that can then serve as the basis of educational efforts in general. Over the decades, the dominance of these ways of describing and studying learning has not faced serious challenge in educational discourse, and in the literature of educational technology (with apparently only the exception of Haugsbakk & Nordkvelle). Certainly, there has been heated debate about which political and curricular ends are to be met through which specific educational means, and which methods of teaching are most efficient under which circumstances, but the language and configuration of learning as a universally occurring process to be facilitated through instructional technique remains uncontested. It effectively delimits the conceptual horizon in which discourse and debate concerning educational technologies and associated practices and policies are confined.

Effects of the Language of "Learning" for Education

By framing learning as a biological or cognitive function that is at base malleable, naturally occurring and universal, psychological learning theories have the effect of stripping away layers of social and institutional significance and connotation, and political, historical, and cultural context. However, when their conceptual implications are followed closely, it can be argued that learning theories go even further than this. They render references to everyday phenomena such as the self, its intentions, agency, or responsibility—even consciousness itself—conceptually problematic. Before concluding, and returning to the matter of the technological nature of learning theory, this chapter will consider a few examples of terms in the favored in the language of learning. It will show how these terms bring quite different implications than some of the alternatives they displace, and how alternative terms are rendered problematic by the quest for universal causal models of learning.

We can begin with the everyday language we use to talk about ourselves, our intentions, choices, and responsibilities. It is important to note that behaviorist learning theories began by banishing this vocabulary as "mentalist." Thorndike, Skinner, and others saw these terms as a mystical or unscientific way of talking about what humans do and why we do it. It is only after Thorndike and Skinner that psychology was able to equip teachers with a developed conceptually consistent vocabulary that allowed them to speak, for example, with some nuance of a self—specifically of an individual's "self-concept," of his or her "self-esteem," "self-efficacy," or "self-image." At the same time as it added these terms to the conceptual vocabulary available to education, cognitivism and its constructivist variants have labeled a wide range of other ways of discussing mental and educational phenomena—experiences such as wanting, remembering, and forgetting—as representative of a novice, "naive," or "folk psychology." This refers to uninformed everyday "theorizing" that is used to "explai[n] human behavior" not in terms of computational devices and instrumental processes, but through reference to "beliefs, desires, intentions, expectations, preferences, hopes, fears" (Baker, 2001, p. 319). Cognitivist learning theory, and the language of learning it provides, encourages or provides for educators and researchers terms for ready use to speak, for example, of learners' self-efficacy—how they can increase their efficiency through assessments of their own abilities. At the same time, it relegates the language of intentions, hopes, and fears to the realm of folk wisdom or naiveté—despite the obvious educational need to ask students or teachers, for example, about their responsibility to others, or of their intentions, hopes, or fears. With their "goal of providing a sound scientific foundation for education" (Sawyer, 2000, p. 15) and their integration of cognitivist, sociocultural, and neurological approaches, the new learning sciences have largely reproduced this vocabulary in their contributions to theories of learning.

It not difficult to see that there is a similar systematic consistency and far-reaching consequence in the way in which very basic terms such as learning, learner, and learning environment have been foregrounded (from the time of Thorndike onward) at the expense of the alternate lexicon of, say, study, student, or classroom. Speaking of a learner, rather than a pupil or student (or in other contexts, a scholar or apprentice), is to imply that this person's role and significance is no longer defined in terms of an institutional and broader cultural context. In their place, the term "learner" elevates a single activity or function—namely of learning—and underscores the importance of its optimization. A similar result is obtained for the teacher, who is either "rendered obsolete" or whose function is otherwise subsumed to the process of learning and the function of the learner. The teacher becomes an enabler, facilitator, or figurative midwife to a learning process that would continue to occur regardless (naturally or innately, but perhaps less efficiently) without his or her presence. The associations of the term "teacher" and the roles with which it is broadly synonymous (such as lecturer, pedagogue, or professor) are replaced by ones subordinate to learning, such as a learning facilitator or learning coach.

Finally, the institutional setting of education and the implications brought with it are either ignored or derided within these traditions, which tend to position learning as happening outside of a given institutional context. In these terms, what is important is not the shaping of learning by historical tradition, multiple stakeholders, or its constitution as a social or cultural reality, but how the "environment" causally impacts the elementary constituents of learning. Are these environments sufficiently information rich? Do they provide reinforcement for desirable behavior? In this case also, multidimensional sociocultural configurations are reduced to causal instrumentality as an environment of stimuli or data. As hermeneutician Hans-Georg Gadamer points out, "environment" is significantly not a term that is able to accommodate the complexity of human intentionality, history, and language. A cage or a petri dish can be an environment for a pigeon or a microorganism, but a place with histories, purposes, and language(s), Gadamer reminds us, is more appropriately characterized as a *world*:

> To rise above the pressures of what impinges on us from the world means to have language and to have "world..." The concept of world is thus opposed to the concept of environment, which all living beings in the world possess...In a broad sense...this concept [of environment] can be used to comprehend all the conditions on which a living creature depends. [Humans are not simply] embedded in their environment. (Gadamer, 2004, pp. 440–441)

Gadamer is making the case for a particularly *human* understanding of experience and learning as culturally and socially mediated. In particular, he is saying that whereas (some) animals only inhabit an environment or habitat, a set of conditions on which they are causally *dependent*, humans exist in a world, a meaningful social, cultural situation in which they *participate*. Insofar as education is a matter of harnessing animal learning, or of mobilizing computational processing or representation, it is removed these humanistic terms. Whether it is referred to as posthuman or simply as "scientific," Gadamer shows how such a conception of education can be characterized as a contradiction in terms. Stripped of social and cultural attributes—intention, value, purpose, meaning—education simply becomes a synonym for natural-scientific control and manipulation, rather than a phenomenon with histories, human purposes, and meanings.

Of course, in making these arguments, I am neither suggesting that many aspects of human cognition and behavior *cannot* be explained in terms of environmental causality, nor am I arguing that the roles and institutional structures of traditional schooling are to be uncontested, possessing value independent of their history and their political and cultural context. Instead, what I am arguing is that this value is contingent and is to be contested precisely in cultural, political, and historical terms. I am arguing, further, that cultural and social contingency that applies to educational forms and methods should also apply to definitions of human learning and learning environments as well. This can be achieved through an interrogation of the value of learning and its theories. In concluding, I make the case that such contingency is in clear evidence in learning and its theories as well.

Technology, Instrumentality, and Ideology

To frame the social, political, and other contingency of learning, I return to the more general issue of technology. As I have already observed, even the most rudimentary definitions of the word indicate that technology reaches far beyond mere artifacts and devices to include processes, methods, means, and applied knowledge generally. Through the historical overview I have provided, I have illustrated how closely interwoven technology as contemporaneous device and artifact is with technology as applied knowledge and method. But their interconnection is not merely one of mutual affinity. Both technology and theories of learning frequently present themselves under the guise of universal instrumentality and natural necessity, whereas in fact, they are both actually and inextricably social and constructed. As I have shown, to speak of "education" in terms of the facilitation of biologically based processes is to strip the historical and cultural context to an elementary paradigmatic scene of learning, an animal in a cage receiving rewards and punishment, an infant in a crib processing the linguistic data available in the environment, or the distribution of cognitive or representational processing between mind and machine. To theorize learning in these terms is also to attempt to establish a foundation beyond culture and human construction from which a broad range of means-ends for relations and efficiencies would incontestably follow. The use of mechanical and digital technological devices in education is merely one of these apparent means or efficiencies, but given the attention and funding these educational technologies attract, it is not an insignificant one.

Beginning with the Frankfurt school, critical theory has seen technology, in material and other forms, as fundamentally ambivalent. Technology is regarded neither as a pure expression of a universal, scientific, and "instrumental rationality," nor as merely the expression of dominant interests and ideology. Andrew Feenberg uses Herbert Marcuse's phrase, "technical rationality," to characterize this fundamental ambivalence:

The dominant form of technological rationality is neither an ideology (a discursive expression of class interest) nor is it a neutral reflection of natural laws. Rather, it stands at the intersection between ideology and technique where the two come together to control human beings and resources. (Feenberg, 2002, p. 15)

Theories of learning, like technological artifacts, stand at the intersection of technique (or science) and ideology (or social and political contingency). Indeed, given the significant changes in learning theories in the last few decades alone, it appears that these theories are allied more closely with ideology than with any neutral, natural, unchanging laws. Like technical artifacts, educational theories, and organizations or instructional design techniques illustrate quite clearly that as technology, they are not just "instrumental to a specific goal" but rather, they "shape a way of life"—as Feenberg (2011, p. 67) puts it. School provides a way of life for the children that attend it, and often the basis for a wide range of related intergenerational and communal practices. Typical instructional design procedures provide ways in which the professional life of teachers, educational technologists, and many others are organized and rationalized. Such forms of organization and technology, of course, are not ambivalent simply in the sense that they may serve positive or negative ends, or that they are not intrinsically good or bad. Their ambivalence arises from the way they frame and shape experience, and how they define educational priorities, problems, and their solutions. Instructional design, like instructional technology, shapes experience and also favors certain uses (with their own political and social connotations) over others, rather than simply serving as a neutral means to a given end. But it can be argued that providing a purely rational, instrumental account of education, as the language of learning has sought to do, does not in any way exhaust the significance of education as a process, product, or experience. Indeed, providing just a rationalist account reifies and mystifies these phenomena, turning them into ideology.

Education, it can be argued, is so inextricably enmeshed with cultural and social significance that to conceive of it in terms of the provision of a kind of hothouse environment for the optimal growth or operation of a specific organ or process is to reduce it to a functionalistic caricature. One can argue, for example, that it is in the practices and contexts of education (rather the construction of knowledge, processing of data, or conditioning of behavior) that knowledge is itself both transmitted and formed in society. Indeed, sociologists of knowledge like Durkheim, Weber, and Foucault have provided detailed and sophisticated accounts that explain these social and cultural dynamics of the construction and reproduction of knowledge and the control or shaping of behavior. As these accounts show, the generation, transmission, and modification of knowledge does not occur through the exercise of innate abilities or universal processes. Instead, it occurs through structures and processes that are by their very constitution indisputably social, political, and ideological—religious practice, political hierarchy, cultural tradition—that themselves shape and are shaped by what is generated, transmitted, and preserved. To say that these structures and processes can be explained, sustained, and driven by an unchanging biological learning process would be to confuse cause with effect, to fail to account for the remarkable and changing variety of social and cultural realities associated with education.

It can also be argued that the ends that educational institutions serve are multiple and inconsistent. For example, education is not solely or even primarily about preparing children for the twenty-first-century workplace or economy. The many moral and political controversies surrounding the school, classroom, and teaching attest to the fact that it is (at least in the West) closely related to the bourgeois construction of the family, to the reproduction of personal and traditional beliefs and values carried by the family and the community, and to the propagation of beliefs and values more generally in society. Education is accordingly rife with ritual, ceremony, and tradition (e.g., entrance and exit rituals like the first day of school and graduation), however much these may be described in rationalized terms of imprinting, punishment, and reward. Education, in other words, encompasses nothing less than the reproduction and (hopefully) also the modification of the existing social order, with the all-encompassing list of stakeholders and roles that this would imply. In this context, means and ends are difficult if not impossible to separate. Is the education of children and youth—and in this sense, social reproduction generally—the expression of a larger goal or purpose, or is it an end in itself? There are no easy answers, except to reject the claim that education (and the technologies used within education) can or should be rationalized and organized around principles vainly aspiring to transcend such underlying questions.

Note

1. It is important to note that Vygotsky also makes it clear that he sees learning as much more profoundly social and cultural than such an interpretation would suggest. Along with science itself, education is inescapably, foundationally or "always already" social and cultural, "voluntary attention...logical memory, and...the formation of concepts" are all first developed "on the social level and [only] later, on the individual level" (Vygotsky, 1978). Vygotsky's and others' insistence that knowledge is always already social and cultural will later serve as an important point in this chapter's conclusion. Here is a more extensive characterization from Vygotsky. In learning, an "interpersonal process is

transformed into an intrapersonal one. Every function in the child's cultural development appears twice, first, on the social level, and later, on the individual level; first, between people..., and then inside the child. This applies equally to voluntary attention, to logical memory, and to the formation of concepts. All the higher [mental] functions originate as actual relations between human individuals" (1978, p. 57). In this sense, Vygotsky can be read as saying that the social is the absolute foundation of the human knowledge (including scientific knowledge), and that there is no causal or computational account of learning that would precede or undergird it.

CHAPTER 3

Networked Cosmopolitanism? Shaping Learners by Remaking the Curriculum of the Future

Ben Williamson

Introduction

What is the future of the school curriculum? Although the use of digital technologies has proliferated widely in schools, their influence on how school curricula are conceived and developed is profound yet often overlooked. In this chapter I argue that curriculum development and design is now increasingly shaped by "the problems technology poses, with the potential it promises, and with the models of social and political order it seems to make available" (Barry, 2001, p. 2). This chapter explores prototype designs for the "curriculum of the future" (Young, 1998) and it interrogates the ideas about the future social and political order they embody, and the ideas about learning and learner identity they promote. The two examples are "Quest to Learn" (Q2L), a "high school for digital kids" opened in New York City in 2009 "where students learn to see the world as composed of many different kinds of systems," and "Learning Futures," a UK school transformation program, which repositions school as a "learning commons" in a web of "extended learning relationships."

Q2L and Learning Futures embody a new fusion of ideas about technology and knowledge within the curriculum. I use the neologism "epistechnical systems" to refer to the binding of technologies and knowledges into hybrid curricular configurations. Epistechnical systems, like all technological systems, are both *socially shaped* and *socially shaping*. As the products of intentional design processes, they are socially constructed and historically contingent, while as technical products or artifacts they function and act to influence and shape thought and action (Latour, 1987; Bijker & Law, 1992; Monahan, 2005). This chapter is concerned with the politics embodied in and catalyzed by the epistechnical systems of new curriculum programs. It asks what authority and what expertise has contributed to the design of these prototype curriculum projects, what politics and values have galvanized their design processes, and it seeks to understand what kinds of prospective identities, actions, and forms of "learning" are to be shaped and sculpted through such systems.

The chapter interrogates the formation of what I call "networked cosmopolitanism," a way of thinking about the future that is infused with normative ideas about the cosmopolitan potential of networks. I unpack the concept of the network as it has become embedded in contemporary thought, and explore its interconnections with the normative ideals of consensus, harmony, empowerment, autonomy, and emancipation associated with the concept of cosmopolitanism. In foregrounding the networked cosmopolitan aspects of recent thought about the future of the social, political, and technological order, my point is to show that curriculum-making processes are thoroughly interconnected with other things, theories, concepts, practices, organizations, structures, and institutions (Fenwick & Edwards, 2010). The chapter then provides some grounded examples of networked cosmopolitan thinking in curriculum reform. Specifically, Q2L and Learning Futures are interwoven with a way of thinking about the future social and political order that is preoccupied with the cosmopolitan potential of networks. As socially shaped and socially shaping systems, these curriculum projects are thoroughly constitutive of this style of thought.

From a curriculum theory perspective, the study of curriculum reform is important because the curriculum is a microcosm of the wider society in which schooling takes place. It constitutes the "intellectual centering" of schooling (Pinar, 2004), and acts as a "message system" designating what society has decided to pass on intergenerationally as "official knowledge" and "real culture" (Apple, 2000; Bernstein, 2000). The curriculum is historically contingent, aligned with fluctuating preferred visions of the future of society, and its organization, content, and form are understood as playing a part in enacting changes to society (Scott, 2008). Curriculum reform, in other words, is political business, always grounded in systemic trajectories of influence that are linked longitudinally from the past into the present and from there projected into the future (Goodson, 2005). The prototypical curriculum developments of Q2L and Learning Futures offer the same basic systemic narrative that constantly changing technologies, accompanied by long waves of social change in all economic, political, and cultural dimensions of existence, have contributed to the need for curriculum reform. These curriculum programs constitute new ways of thinking about the future of schooling that construct globalization as a rationale and a set of "imperatives" for educational reform (Ball, 2008; Rizvi & Lingard, 2010). Such programs have made the curriculum into a problem for resolution in a contemporary period that is increasingly described as being globalized, multipolar, cosmopolitan, and, above all, digitally networked (Castells, 1996; Held, 2010). Moreover, these notions of networks assume technical change to be the model for appropriate contemporary curricular invention. As Barry (2001, p. 87) states, "Networks do not so much reflect social, political and technological reality; they provide a diagram on the basis of which reality might be refashioned and reimagined: they are models of the political future." In the same way, curriculum projects including Q2L and Learning Futures adopt the network diagram as a model for reimagining and refashioning the future of the school curriculum.

Minor Reforms and Microcosmic Futures

The chapter is based on research that has sought to trace the participation of cross-sectoral organizations and alliances in contemporary curriculum reform. The data consist of textual documents from a range of curriculum reform programs, including reports, curriculum and pedagogical guidance, manifestoes, pamphlets, articles, and essays as well as websites, infographics, diagrams, interactive devices, and other multimedia. These "inscriptions" are material techniques of a particular style of thinking. A style of thought, as described by Rose (2007), is a particular way of thinking, seeing, and practicing in a particular field of inquiry. It puts frames and limits on what counts as an argument or an explanation. It is underpinned by key terms, concepts, references, and relations, and is linked to key techniques of intervention. But a style of thought is not merely explanatory. It actually shapes and establishes the problems, difficulties, and issues that an explanation is required for. It modifies or remakes the very things it explains. Focusing mainly on two case study examples—Q2L and Learning Futures—I examine how their ideas are created from fragments, slogans, and recitations that are reiterated over time by a variety of actors within public education (Ball, 1994, 2007). These texts are made up from the juxtaposition of "bits and pieces" (Law, 1992) into "inscription devices," which stabilize myriad different ideas, concepts, theories, practices, designs, and their historical and political networks of connections, into a new discourse, conceptual vocabulary, and a set of practical possibilities for thought and action on the curriculum (Fenwick & Edwards, 2010).

The research recognizes, then, that any educational program is complex, contingent, and heterogeneous. Rose (1999b) refers to the functional networks that constitute educational programs and interventions as "the technology of schooling." A technology of schooling consists of pedagogic knowledges, educational theories, classroom organization, timetables, techniques of instruction, supervisory regimes, behavioral and regulatory codes, curricular guidance, inscriptions, and digital devices, among other things, all brought together through diverse associations and delicate affiliations between bit-part players, actors, and agencies with the aim of managing students' capacities and habits. It is "not implanted through the monotonous implementation of a hegemonic 'will to govern'" but is instead "hybrid, heterogeneous, traversed by a variety of programmatic aspirations" (Rose, 1999b, p. 54).

The metaphor "centrifugal schooling" describes a loose alliance of prototype curriculum programs that have emerged recently in the United Kingdom, Australia, and North America (Williamson, 2012). As a metaphor, centrifugally registers the increasing decentralization of curriculum development. Rather than being dictated by centralized education systems, such curriculum projects are increasingly being thought, influenced, and made in all sorts of "local centers." Centrifugal schooling consists of many separate innovations and curriculum programs generated and promoted by a loose alliance of new curriculum gatekeepers, brokers, entrepreneurs, fixers, catalysts, intellectual workers, and innovators from think tanks, nonprofit organizations, charities, and the philanthropic outgrowths of corporations, and from there transmitted out into schools and pedagogic practice. Some have come to nothing, failed, or abandoned; others have extended, been linked up to other sites and other inventions, coupled up with other organizations, and linked into other networks of resources, and have become established as stable and lasting networks of thought and action.

Rather than top-level curriculum reforms mandated by state power, the focus for studying centrifugal schooling is on minor-level *reforms-in-action*, prototypical curriculum developments, and the microlevel experts and authorities that act as a loose relay between "certain administrative and political practices and a diversity of local initiatives" (Jensen & Lauritsen, 2005, p. 365). Consequently, these programs do not necessarily produce a clear and coherent vision of the future to which they point but a complex, messy, contradictory, confused, and unclear set of seemingly contrary alternatives and dilemmas (Ball, 2007). Such programs act as minor, distant, and micropolitical relays of political aspirations and objectives for the future thoughts and actions of the young.

Ultimately, the design of the curriculum of the future is a matter of molding students' "prospective identities" to "*deal with cultural, economic and technological change*" (Bernstein, 2000, p. 67; emphasis in original). Prospective identities are set on the path to a "new future" by "gatekeepers and licensers" (p. 76) who translate images of the future into legitimate pedagogic practice. The future-facing prospective identities promoted in Learning Futures and Q2L, I argue, are shaped by new kinds of curriculum gatekeepers through an amalgam of cosmopolitan principles of autonomy and self-responsibility and network notions of connectivity in order to motivate a particular style of belonging in the (imagined) future of a globalized society.

A network-conscious style of thinking about the curriculum, then, is far from socially neutral or nonpolitical. It shapes and prefigures possible curricular futures. As Gough (2002) argues, curricula are not just "out there" waiting to be discovered, but must be imagined and constructed. Curriculum reform processes are not straightforward microcosms of a social reality that already exists beyond the school, but microcosms of *imagined* social futures—microcosmic futures-in-the-making. By looking at how programs such as Q2L and Learning Futures are constructed, and by identifying the style of thinking that galvanizes them, we can glimpse something of new futures being imagined for young people. They construct consensus and legitimacy around the construction of particular problems to which they also offer solutions. The focus here is on the problems and objectives they frame, the visions and futures they catalyze and foreclose, the curricular solutions they propose and promote, and the politics embodied in their style of thought.

Network Consciousness

Networks have become integral to a contemporary style of thinking and imagining a whole new social and political order. In comparison to the twentieth-century industrial era of mass production, centralization, and organized hierarchy, pinpointed by the image of a single central dot to which all strands led, the twenty-first-century digital age has been defined by the "death of the centre" and its replacement by a mesh of many points all linked multidirectionally into webs and networks (Ryan, 2010). Systems, complexity, feedback, matrices, lateral connections, associations, openness, hybridity, fluidity, multidimensionality, and connectivity are the new cultural keywords in a smart world of networks where the dynamic and the mobile are challenging centralized bureaucracy, dialogue and cooperation are preferred to hierarchical authority and order, horizontality has conquered verticality, flexibility seems more important than routine, and a counterculture of the high-tech "geek" has taken over from the dark-suited manager of the big firm (Žižek, 2008). The "network society" (Castells, 1996) is a lateral society of fluid networks rather than a vertical society of totalizing structures (Bauman, 2007). The figure of the network has become the dominant sociotechnical diagrammatic form for many contemporary problems and projects (Barry, 2001).

Unsurprisingly, the potential for network-based technologies to introduce new interaction, dynamics, and participation into everyday life and publics has become a powerful unifying belief for many educational technologists (Ito et al., 2010; Boyd, 2011). Learning is now increasingly understood within this field to be decentered and dispersed in time and space, horizontally structured, taking place fluidly throughout lifetimes, networked and convergent across many different media and locations, with the internet itself imagined as a learning institution in an increasingly "open era" (Davidson & Goldberg, 2010). As educational critics have shown, the hyperconnected ideal of a radically open "post-school era" (Facer & Green, 2007) reanimates the countercultural "deschooling" agenda as a convivial digital curriculum which facilitates communication, cooperation, caring and sharing between free agents, and distributes learning into a nomadic network of authentic practices, cultural locations, and online spaces (Suoranta & Vadén, 2010). This is, it seems, a kind of networked neoprogressive utopia, a high-tech deschooling of society, "leaving us all enmeshed in Illichian webs and nets" (Hartley, 1997, p. 155). Such images are set against schools caricatured as innately conservative institutions that continue to rely on structured hierarchical relationships, a static print culture, and old-style transmission and broadcast pedagogies, which are at odds with the networked era of interactivity and hypertextuality (Selwyn, 2011a). Indeed, networks have become a seemingly institutionalized utopia with unlimited ameliorative potential for education, despite evidence of their negative capacity to catalyze disunity, disconnection, and dysfunctionality, and to reduce educational knowledge to marketable commodities, "soundbites," and populist user-generated knowledge (Frankham, 2006; Ferguson & Seddon, 2007; de Lima, 2010).

Moreover, the optimistic rhetoric of networks glosses over the complexity of networks in their cultural context. Castells (2009) has documented the emergence of a new dominant "cultural pattern" in the age of the internet— networked individualism: the construction of individual cultural worlds in terms of personal preferences and projects. Networked individualism is a culture that starts with the values and projects of the individual who interacts with others following their own choices, values, and interests, rather than by tradition and hierarchy:

> The culture of networked individualism finds its platform of choice in the diverse universe of mass self-communication: the internet, wireless communication, online games, and digital networks of cultural production, remixing and distribution... [T]he culture of networked individualism can find its best form of expression in a communication system characterized by autonomy, horizontal networking, interactivity, and the recombination of content under the initiative of the individual and his/her networks. (Castells, 2009, p. 125)

Networked individualism resonates with the "culture of freedom," autonomy and experimentation associated with the Silicon Valley "culture of hackers" and the "culture of the designers of the internet" itself (Castells, 2009, p. 125). The seemingly "liberating potential of networks" represented by this culture of freedom, however, has largely been enclosed and expropriated "to expand for-profit entertainment and to commodify personal freedom" (Castells, 2009, p. 414).

Networked individualism is a style of thinking derived from a "cyberlibertarian" politics of individual freedom, choice, and entrepreneurship that now finds its way into the field of education as networked learning. The network consciousness of individual freedom has been legitimized through the expertise of educational technologists to shape the thoughts and actions of the young, although many technologists (e.g., Lanier, 2010) and educational specialists working in new technology research (e.g., Buckingham, 2006; Selwyn, 2011b) have expressed alarm about the uncritical acceptance of cyberlibertarian values of radical individualism and self-interest in much educational technology research. It is therefore important to acknowledge the political and indeed human implications of a networked world in which "individuals are becoming more active, their capacities to know and to question enhanced by transnational media, the internet and many other communicative technologies" (Miller & Rose, 2008, p. 217). The internet empowers "lay experts" and "experiential experts" whose authority and expertise resides not in training, status, or skills but in experience (Rose, 2007, p. 126). An expertise of personal autonomy and freedom shaped by the internet gives rise to very new understandings of human capacities and identities, and raises difficult questions for education.

Cosmopolitan Combinations

The network consciousness may now be linked up with the production of a modern identity that Popkewitz (2008) has called a "cosmopolitan self." Cosmopolitanism celebrates empowerment, voice, and emancipation from traditional habits and attitudes. Cosmopolitanism is embodied in talk about autonomy, self-responsibility, respect for diversity and difference, and participation and collaboration in communities; and it focuses on the creation of a "good" or "ethical" future. However, Popkewitz (2008, p. 3) is skeptical about these enlightenment ideals and transcendent values in today's "reform society," where "the seductiveness of reform is its promise of cosmopolitan harmony and consensus." Cosmopolitanism is a "mode of life organized in pedagogy" (Popkewitz, 2008, p. 5). Here, Popkewitz describes as cosmopolitanism a "liberal freedom" which is mobile, no longer bound "to a sense of identity built through geographical location and face-to-face interactions," and refers to a "global citizen who has a traveling home" (Popkewitz & Bloch, 2001, p. 89). This is rather like Beck's (2006, pp. 3–5) analysis of the "boundarylessness" and "conditions of cultural mixture," which permits "quasi-cosmopolitan" individuals to construct "a model of one's own identity by dipping freely into the Lego set of globally available identities." Calhoun (2002, p. 893) calls such "actually existing" cosmopolitans "frequent travelers" who take a "good ethical orientation" into "the frequent-flyer lounges."

Translating the theme of cosmopolitanism in education, Popkewitz, Olsson, and Petersson (2006, pp. 432–433) emphasize how the cosmopolitan identity fabricated in the global information society is an autonomous, agential, self-responsible, and empowered lifelong learner who solves problems, has a voice, makes choices, and collaborates in communities of learners through the computer and the internet. Rizvi (2009, pp. 260–261) summarizes such a "corporatist view of cosmopolitanism" as "global connectivity":

> It celebrates individuals who are able to take advantage of global mobility…encourages values that are associated with global economic exchange, social entrepreneurialism and cultural adaptability…facilitated by the unifying potential of global telecommunications and digital information systems to drive a new culture of consumption.

Likewise, Camicia and Franklin (2010), investigating cosmopolitanism in curriculum reform, conclude that there are two competing discourses of cosmopolitanism. One is a discourse of "democratic cosmopolitanism," which defines global citizens as a community that behaves optimally when government regulations support cultural representation, human rights, and social justice. The second, more dominant discourse is "neoliberal cosmopolitanism," which defines global citizens as a community of self-starting entrepreneurs who function best when government regulations support market rationality. This is cosmopolitanism as market decentralization and the optimizing of individual free choice. In short, this is the cosmopolitanism of the culture of mass self-communication associated with networked individualism—or an emergent form of networked cosmopolitanism. As a mode of life organized in pedagogy, networked cosmopolitanism emphasizes personal preferences and projects, global (virtual) mobility, connectivity, autonomy, experimentation, problem solving, and communities of learning. Above all networked cosmopolitanism embodies an ethic of autonomous self-improvement and the assumption of personal responsibility that "promises to make it possible for us all to make a project of our biography, create a style for our lives, shape our everyday existence in terms of an ethic of autonomy" (Rose, 1999a, p. 258).

In the following sections, I illustrate how networked cosmopolitanism is interwoven with the objectives and aspirations of two prototype curriculum projects. This is not to claim such programs are realizations of a dominant corporatist or neoliberal agenda or other hegemonic attempts to enclose the future of education. Rather, curriculum reform is interwoven with this network-conscious and cosmopolitan style of thought—consisting of shared vocabularies, theories, and explanations—which crosses and bisects sectoral, disciplinary, and ideological boundaries. These programs make the future of the curriculum intelligible, thinkable, and practicable in networked and cosmopolitan terms. I detail the various actors, experts, and agencies involved in promoting them, and then finally return to the question of the kind of prospective learner identities they generate.

Complex Systems: The Case of Q2L

Located in New York City and touted as a 'high school for digital kids', Q2L was established in 2009. A collaboration between the nonprofit Institute of Play and the education reform organization New Visions for Public Schools, the Q2L's curriculum and pedagogy emphasizes "design, collaboration, and systems thinking as key literacies of the 21st century" (Salen, Torres, Wolozin, Rufo-Tepper & Shapiro, 2011). A sister school was established in Chicago in 2011. Q2L is interesting because it has originated outside the usual organs of the education system. Its key collaborating partner, New Visions for Public Schools, works throughout New York City by creating new small schools and opening its own charter high schools. Its approach to educational improvement involves all stakeholders—the public school system, government, businesses, community groups, parents, and students—working "harder to do better together," and some New Visions reforms have been replicated in other sites as it extends through the education system. Its funding comes from a mix of government, corporations, philanthropic foundations, and individual sources. Q2L originates, then, from a cross-sectoral and interorganizational network of minor actors and agencies whose authority and expertise extends in reach and influence through the multiplicity of their affiliations.

The conceptual and organizational model for the Q2L's curriculum was designed by the Institute of Play, a games and learning nonprofit organization staffed by professional game designers and researchers in the field of game-based pedagogy, new media literacy, and the learning sciences. The Institute of Play's director is also the director of Q2L and a well-respected researcher and developer in the field of videogames and learning. Additional support on Q2L has come from Parsons School for Design, particularly its mixed-reality lab, and it has consulted with curriculum and teaching experts, middle school students, and selected academic experts involved in researching digital media and learning. Rather than a unique outpost of innovation, Q2L is located in a matrix of interorganizational and cross-sectoral relationships and reformatory networks involving various kinds of actors, authorities, and experts, each with their own ideas and historical, conceptual, and political associations.

Moreover, Q2L is a key project in a major US program to promote "connected learning" (see connectedlearning.tv). According to the director of this program, connected learning

> harnesses the powerful new connection to ideas, knowledge, expertise, culture, friends, peers and mentors we have through the internet, digital media and social networking... Connected learning is an answer to three key shifts as society evolves from the industrial age of the 20th century and its one-size-fits-all factory approach to educating youth to a 21st century networked society:
>
> 1. **A shift from education to learning.** Education is what institutions do, learning is what people do. Digital media enable learning anywhere, anytime; formal learning must also be mobile and just-in-time.
> 2. **A shift from consumption of information to participatory learning.** Learning happens best when it is rich in social connections, especially when it is peer-based and organized around learners' interests, enabling them to create as well as consume information.

3. **A shift from institutions to networks.** In the digital age, the fundamental operating and delivery systems are networks, not institutions
such as schools, which are one node of many in a young person's network of learning opportunities. People learn across institutions, so an
entire learning network must be supported. (Yowell, 2012, n.p.)

The model of connected learning is based on an epochal narrative that a
systemic trajectory of change—the formation of a network society—has
brought about an imperative for educational reform.

Through its network of experts and its indirect affiliations, Q2L offers a
blueprint for a possible future of institutional schooling in the network society, one where the school is but one node in the learning networks of each
of its students. The school's main documents emphasize "systems thinking"
and "learning about the world as a set of interconnected systems" and it is
"committed to graduating strong, engaged, literate citizens of a globally networked world" (Salen et al., 2011). To act in such a world, students need to
be able to recognize patterns and identify structures, think connectively and
creatively, be inventive and innovative, adopt and tolerate multiple cultural
perspectives, exhibit empathy and reciprocation, understand what it means
to be an active global citizen, understand and respect self and others, and
understand the various modes of new media communication.

Q2L also has a strong emancipatory ethos. It positions its students as
"sociotechnical engineers" who can create systems (games, models, simulations, stories). By "designing play," it claims, "students learn to think analytically, and holistically, to experiment and test out theories, and to consider
other people as part of the systems they create and inhabit." The inbuilt
creativity and design focus seeks to produce students who are empowered
to act and make and participate in global dynamics rather than receive and
consume. In order to do so, Q2L also provides a structured curriculum
model, which claims to "interweave state standards with ways of knowing
and doing." The curriculum is organized as interdisciplinary knowledge
domains instead of separate subjects. Each interdisciplinary domain provides experience in "integrated expertise" such as "researching, theorizing
about, demonstrating, and revising new knowledge about the world and the
systems of which it is composed."

The integrated domains are described as follows. "The way things work"
integrates science and math and involves "taking different kinds of systems
apart and modifying, remixing, and inventing systems." In the "being, space
and place" domain, students study "time, space and human geographies as
forces that shape the development of ideas, expression and values" through
combinations of social sciences and English language arts. "Codeworlds"

blends language arts and math and computer programming and involves students "decoding, authoring, manipulating and unlocking meaning" through the "interpretation of symbolic codes ordering our world." "Wellness" situates personal, social, emotional, and physical health within systems of peer groups, family, community, and society. Finally, "sports for the mind" emphasizes the "fluent use of new media across networks" for a "productive career, prosperous life and civic engagement in the 21st century." The interdisciplinary curriculum is delivered through problem-based "missions," "levels," and "quests," which are organized according to basic videogame architecture.

Based on this strong "networks and systems" style of thinking, Q2L reimagines "school as just one kind of learning space within a network of connected learning spaces that spans in school, out of school, local and global, physical and digital, teacher led and peer driven, individual and collaborative." Indeed, here learning itself is understood and represented as a complex system that bisects disciplinary boundaries and the spatial and temporal constraints of traditional schools. Systems thinking refers to the understanding that any system—social, technological, natural—maintains its existence and functions through the dynamic interaction and interdependence of its parts and stresses the unintended consequences of complex interactions and relationships. It is antithetical to the traditional curriculum of insulated subjects, isolated facts, and knowledge learned out of context. Q2L "posits learning as context-based processes mediated by social experiences and technological tools," a "highly social endeavor" that takes place through "situated practices" within "communities of practice." In this way, a situated-learning view stipulates that learning cannot be computed solely in the head but rather is realized as a result of the interactivity of a dynamic system. These systems construct paradigms in which meaning is produced as a result of humans' social nature and their relationships with the material world of symbols, culture, and historical elements. The structures, then, that define situated learning and inquiry are concerned with the interactivity of these elements, not with systems in the individual mind.

Through this approach, students at Q2L are engaged in situated and authentic, "real-world learning experiences." The Q2L's distinct conceptual framework for the curriculum hybridizes the systems language of videogames design with the systems language of situated cognition.

Additionally, Q2L's curriculum for the future represents the world in terms of complex open systems. Q2L's version of complexity theory emphasizes emergence, nonlinear dynamics, uncertainty, feedback loops, self-organization, and interconnection. In complexity terms, learning, curriculum, and knowledge at Q2L are understood as continuous invention and exploration, produced through complex interactions among people, action and interaction, objects, and structural dynamics, which all produce emergent new possibilities. The complexity approach inscribed by Q2L treats curriculum not as a product for imposition but as a process of emergence and interaction. It is forward looking in that it embraces the contingency and uncertainty of educational outcomes. It recognizes processes of inquiry, exploration, and mobilizes a vocabulary of networked interactions and webbed learning. The curriculum, viewed from a complexity perspective, is an open system of constant flux and complex interactions rather than a closed system of prescriptions and linear progressions. A complexity curriculum emphasizes students as knowledge producers, organizing and constructing knowledge as they interact through webbed networks of connections and interconnections (see Doll, 2008).

The Q2L's integrated curriculum is assembled from a mix of open systems and complexity theories into an emergent form of networked, collaborative, and digital interdisciplinarity. Its keywords are "systems," "dynamics," "integration," and "hybridization," and it seeks to prepare students for a world, which it characterizes as globally connected and complex. In these terms, it can be seen as a hyperconnected curriculum of the future. Despite the high-tech, digital interdisciplinarity discourse of game design, however, it is also constituted by a more affective, emotional, and ethical discourse that resonates with a much longer curricular legacy in the United States. The basic intellectual architecture is derived from John Dewey's emphasis on "inquiry," "experience," and "learning community," as refashioned through the discourse of open systems and networks. It amalgamates participation in the economic sphere with notions of community and local responsibilities in the cultural sphere. The first is promoted through emphasizing technological competence and the soft skills required for flexible working; the latter through appealing to authentic and learner-centered or "personalized" learning. It offers a hybrid language of learning that is not only high-tech but also emotionally "high-touch" (Hartley, 2006). The technology of the Q2L's curriculum is an assemblage formed by the folding together, juxtaposition, and hybridization of these seemingly diverse elements, theories, histories, and technologies. These elements have been brought together and "lashed up," ordered, and assembled by a plurality of actors as a temporary stabilization of knowledges and practices through a "multiplicity of curriculum-making practices" (Fenwick & Edwards, 2010, p. 58).

Expert Learning: The Case of Learning Futures

A second example of messy curriculum making among diverse actors is Learning Futures. Learning Futures aims to support students to "work and thrive as the world grows more interconnected, the environment becomes less stable, and technology continues to alter relationships to information" (Price, 2011, n.p.). Established in 2008 by the nonprofit Innovation Unit and the philanthropic Paul Hamlyn Foundation in the United Kingdom, Learning Futures has worked with 40 schools to develop innovative changes to curricula, pedagogy, and assessment. In this section, I first want to describe something of the network of actors involved in assembling Learning Futures, and then to examine the curricular objectives and pedagogic principles it mobilizes.

The organization behind Learning Futures, the Innovation Unit, originated within the UK government department of education as a source for innovative ideas in public service delivery. It was an early adopter of slogans of innovation in educational reform. According to its publications, the kind of innovations required for the future of education are to be found in open-source hacker communities and in the smart strategies and rapid R&D culture of Silicon Valley (Bentley & Gillinson, 2007). These kinds of self-consciously "iconoclastic" ideas are now synonymous with the "shock tactics" of think tanks in British politics (Mulgan, 2006, pp. 151–152). Peopled by "intellectual workers" with "propellant" and "vehicular" ideas, think tanks have championed small-scale creative innovations in public services (McLennan, 2004; Osborne, 2004). Vehicular expertise is not concerned with the grand schemes of big legislation but with practical, usable, marketable ideas capable of arousing attention and propelling the "buzz" of creativity and innovation. Such expertise contributes to a constantly mobile, creative culture of new ideas, innovations, and intellectual creativity.

Learning Futures talks of inquiry in innovative terms as research, experimentation, learning through doing, problem solving, and evaluating information. One of its main documents makes direct reference to the US-based "Partnership for 21st Century Skills," from which it cites the need to equip students with new skills of collaboration, information literacy, and adaptability—a virtually global standard for innovation. Moreover, its project-based learning approach has been developed directly from a partnership with High Tech High, the network of San Diego charter schools assembled by the Economic Development Corporation and the Business Roundtable, to discuss the challenges of preparing individuals for the high-tech workforce (see hightechhigh.org), placing it clearly in a globalized network of curricular innovation catalyzed by a high-tech reformatory style of thinking.

These connections notwithstanding, Learning Futures encloses its innovative style of thinking in a more affective language of "engaging learners" and "engaging schools." The booklet produced by the Learning Futures' program in collaboration with High Tech High speaks of learning being "passion-led," "fun," "exciting," "inspiring"—it should have "real world" relevance, stretch students' "intellectual muscles" as "expert learners," and "ignite students' imaginations." Learning Futures strongly promotes a thematic and "project-based" pedagogy, which involves "designing, planning and carrying out an extended project" using "digital technology" to "conduct serious research, produce high-quality work," and to "foster a wide range of skills (such as time management, collaboration, and problem-solving)" (Patton, 2012).

The Learning Futures' approach to curriculum is defined in terms of being "placed," "purposeful," and "pervasive." It reaches, has relevance to, and connects with students' own communities and interests; fosters value and agency and encourages students to behave as "protoprofessionals." It also extends into independent and interdependent informal learning that "matters to students" (Price, 2011). The task for teachers in an inquiry classroom is to listen and respond to students, adapting flexibly and fluidly to their interests and questions accordingly. These examples of student-led inquiry, personal projects, and portfolios constitute what Fendler (2001, pp. 132–133) has called "interactive pedagogy":

> In interactive pedagogy...the teacher teaches by adapting the material to the child's momentary interests and imparts information that is set by the children's questions. This pedagogy requires the teacher to respond flexibly to the child's feelings, words, and actions... [I]nteractionism constructs both a response-able/-ready child and a response-able/-ready t eacher... Interactionism...can be characterized as fluid, dynamic, situation responsive, pragmatic and virtual.

Inquiry learning and interactionist pedagogies are mutually interdependent. They privilege the idea of students as active constructors of knowledge through a variety of collaborations sourced through the web. These approaches have opened the curriculum to a diverse array of new providers and new classifications of knowledge, new sources of expertise, and new pedagogies crafted around the "know-how" of competence rather than the "know-what" of curricular knowledge. The student constructed by Learning Futures is an active, autonomous, inquiring, and constructivist learner.

In its efforts to generate these new constructivist learning identities, the Learning Futures' approach to curriculum, like Q2L, is informed by a scientific expertise of learning rather than by curriculum theory with its concern with the politics of knowledge. Biesta (2006, pp. 16–17) describes the emergence of a "new language of learning" that is assembled from a composite of constructivist and sociocultural theories of active knowledge construction, an increasing consumer economy of lifelong learning, increased emphasis on generic learning outcomes, and a narrowly individualistic and psychological view of the learner (see Friesen, this volume). Learning science amalgamates psychological, cognitive, sociocultural, and increasingly neuroscientific subdisciplines, along with computer science and engineering. Selwyn (2010, p. 67) notes that "a 'learning science' perspective on educational technology now pays close attention to the technical and the social processes of learning with digital technology," and to "the perceived technological and psychological strengths and shortcomings of individual learners, their tutors, and educational institutions," but it is far less concerned with "the wider social contexts that make up education and society." The rise of "learning science" has been mirrored by the decline of curriculum theory and the concomitant depoliticization of educational discourse.

Taking a slightly different perspective, I understand the theories and values of learning science itself to be playing a part in the social shaping of learners. Learning science is not merely explanatory of learning, but actively shapes the ways in which learning is understood and the ways in which students are encouraged to understand themselves as learners. Learning science functions through scientific expertise generated from various psychological subdisciplines to act upon students' identities, thoughts, and actions. That is to say that the learning sciences themselves constitute the wider social context for analysis. The learning sciences are part of the modern technology of schooling, which privileges and generates particular styles of human activity, particular ways of thinking and acting that are designated as appropriate according to the scientific expertise of learning.

Elements of the new depoliticized scientific expertise and language of learning can be found throughout promotional documents from the Learning Futures' program. Students are to become expert learners who understand themselves through the new scientific language of learning. Specific references are made to the "nature of learning," "thinking skills," the "psychology of optimal experience," "learning processes, settings and styles," "peer tutors," "learning communities," "effective lifelong learning," "virtual learning," and "visible learning," all of which are proposed to constitute "learning which is deep, authentic and motivational" (Learning Futures, 2010; Price, 2010). The psychological accounts of the new language of learning have catalyzed a host of reformatory aspirations around notions such as inquiry, constructivism, project-based learning, and competence.

The science of learning represented by Learning Futures shares many characteristics with the networked language of learning associated with Q2L. The new language of learning promotes theories and practices of interactionist pedagogical innovations and interactive technologies that together work upon the competence of the learner, remaking the learner as an active and constructivist self.

Psychotechnologies of Schooling

Despite the seemingly depoliticized nature of the new scientific language of learning presented by Learning Futures and Q2L, these active and constructivist prospective identities are to be actively sculpted to secure political objectives. Here it is necessary to broaden back out to slightly wider concerns in curriculum reform. A report for the UK-based National Endowment for Science, Technology and the Arts (NESTA) on the "wider skills" required for twenty-first-century economies is especially significant. It emphasizes the importance of "new smarts," "orientations," "capabilities" and "capacities," "dispositions" to learning, and the "mental and emotional habits of mind" that are required "if innovation is to be effectively developed in young people" (Lucas & Claxton, 2009, p. 4). Another UK report by the Young Foundation specifically recommends dedicating school curriculum time to "honing social and emotional competencies," "thinking creatively, collaboration, empathy and emotional resilience," especially so since "wellbeing and resilience matter to employment and to the economy" (Bacon, Brophy, Mguni, Mulgan & Shandro, 2010, pp. 49–50).

These reports establish the maximization of the subjectivity of learners as a political imperative in an increasingly technological, networked world, particularly in relation to competitiveness in the global knowledge economy. Schools are positioned as responsible for the cultivation and promotion of habits of mind and the emotional capital required for the nation to remain innovative and competitive. The ideal prospective learner for such a prospective future is to be psychologically armed and pedagogically prepared with the social and emotional competencies and skills required for innovation. The expertise of psychology, with its seemingly impartial and depoliticized truth claims about how the human psyche is understood, has been particular powerful in shaping such arguments. Foucault (2008, pp. 229–230), in fact, showed how innovation required "educational investments" in "human capital" and traced such investments in terms of care, affection, cultural stimuli, family life, health, and other "psychological costs." He called this the "ethical-economic-psychological characteristic of capitalism" (Foucault, 2008, p. 231).

Psychological knowledge and expertise, then, is now essential for aligning the aspirations and inner lives of learners with the objectives of government. Through the "gaze of the psychologist" (Rose, 1999a) the characteristics of the modern learner may be analyzed, evaluated, assessed, inventoried, documented, and otherwise inscribed in order that they can then have things done to them or do things to themselves. The reports detailed above show how the knowledgeable management of human psychology in terms of the adduced capacities, emotional habits of mind, well being, happiness, and competences of each individual learner has now been posited as a route to innovation and global competitiveness in a knowledge economy. Human interiority itself has become the subject of the psychotechnology of contemporary schooling.

Developing this analysis further in line with his theories of cosmopolitanism and curriculum reform, Popkewitz (2012) has shown how the school curriculum is organized through "psychological eyes" and psychological concepts that translate idealized and normative ideas about who the learner is or should be. The curriculum is never just knowledge or content but "embodies learning how to see, think, act and feel" (Popkewitz, 2012, p. 177). Crucially, for Popkewitz (2008) the autonomous, agential, self-responsible, and empowered lifelong "learner" who solves problems, has a voice, makes choices, and collaborates in communities of learners through the computer and the internet is a cosmopolitan identity. The objective of contemporary curriculum reform, he argues, is to shape such a learner. The task of learning science is therefore to translate the principles of cosmopolitanism into psychological concepts that can be applied to the characteristics of the learner.

Curriculum reforms like Learning Futures are inscribed in the language of autonomy and self-responsibility that originates in cosmopolitan principles and is now translated and reanimated in a new scientific language of learning. Q2L and Learning Futures both appeal to an expertise of learning science that purports to understand young people psychologically and scientifically as certain kinds of learners in order to activate new kinds of techniques for acting upon them to achieve their objectives. Through the expertise of learning science, they have fabricated an image of the learner who is active, interactive, and autonomous, and participatory, connected, and networked. The mode of thought associated with learning science is embodied in the myriad ways in which schools are now exhorted to evaluate, know, and act upon the psyche of the learner. The experts of learning science act as the gatekeepers and licensers of new futures, which, embedded in the networked cosmopolitan outlook of curriculum reform programs, encourage and promote new prospective identities. The expert learners cultivated by the project-based and inquiry-based interactionist pedagogies of Learning Futures are ideally prospective identities for such a networked cosmopolitan future. That is to say, the psychotechnological techniques of such curriculum reform programs take as their objective the sculpting of an ideally networked, cosmopolitan pedagogic identity.

In sum, networked cosmopolitanism is a future mode of life, a prospective identity, imagined to be organized in the interactionist pedagogies and epistechnical curricula of centrifugal schooling. As a mode of life made up through the expertise of learning science, networked cosmopolitanism is centered on recombinatorial acumen—the ability to remix and mash up content, to constantly manage, edit, and redact one's personal profiles and networks, and to mix and match different cultural and consumer products in order to express a lifestyle that is simultaneously local and global. Personal autonomy is constructed as a digital project or a portfolio. Digital identity management and the ability to edit one's own "remixed and remixable" self has become a "survival skill" (Papacharissi, 2011, p. 317) in a networked cosmopolitan world. Translated as pedagogy, networked cosmopolitanism promotes such a remixable mode of life through distributed learning, communities of practice, and the aerosolization of learning into the very atmosphere of digital culture. It shapes and sculpts a prospective identity that is individualist and self-interested yet also democratic and cosmopolitan in outlook; entrepreneurial and globally mobile yet also socially activist and locally committed; consumerist yet also countercultural; community focused yet also self-fashioning. Centrifugal schooling, as illustrated by the case studies of Q2L and Learning Futures, is the ideal curricular format for networked cosmopolitanism.

Part of a modern preoccupation with the diagram of networks as a model for the future social and political order, centrifugal schooling is amalgamated from cosmopolitan principles of autonomy and self-responsibility, networked individualist notions of personal freedom and connectivity, and bound together by a normative bundle of psychological classifications of the learner, in order to galvanize a particular style of belonging in the future of a globalized society. Yet the case study examples are not pure implementations of a single ideal form of the curriculum, or isolated and radical inventions. Instead they are socially accumulated and improvised as "shared vocabularies, theories and explanations...between agents across space and time—Departments of State, pressure groups, academics, managers, teachers, employees, parents" (Miller & Rose, 2008, pp. 34–35). This chapter has traced something of the formation of those shared vocabularies, theories, and explanations—a networked cosmopolitan style of thought that is now being inscribed and stabilized in the documents, guidance, websites, and other inscription devices of curriculum reform.

Conclusions

According to the visions of the networked curriculum of the future imagined by Q2L and Learning Futures, isolated and insulated educational institutions are now being challenged by a much more pedagogically polygamous range of incidental, noninstitutionalized learning relationships and attachments. Centrifugal schooling is a borderless and boundary-free vision of connected learning and a networked future of school that can travel anywhere, interactively taking in encounters with knowledges from all areas of experience, from any media, and across all disciplines. In this imagined future, new spatialities and temporalities of learning are opened up by the flexing of timetables, the compression of space by real-time digital communication, and the virtual erasure of school walls in a "borderless economy of education from within which student consumers can pick and choose" (Ball, 2008, p. 198).

Q2L and Learning Futures have been assembled through a mishmashing of such networked and cosmopolitan ideas along with a new language of learning derived largely from psychological accounts of the science of learning. Within this complex of ideas, theories, and explanations, Q2L and Learning Futures can be understood as promoting cosmopolitan identities through the psychotechnologies of schooling; shaping and normalizing new modes of thought, action, and feeling, new prospective identities, in order to achieve the seemingly distant objectives of global innovation in the knowledge economy. They fold into the language and techniques of learning new understandings of learners as participatory, networked, and connected to ideas, knowledge, expertise, culture, friends, peers, and mentors through the internet, digital media, and social networking.

The "mashed up" prospective identities (Fenwick & Edwards, 2010, p. 168) being constructed through the psychotechnological curriculum of the future are, then, made up from a cluster of ideas about the ideal individual for such an age. This is an individual who is globally networked and cosmopolitan, interactivist and constructivist, active and autonomous, freely choosing, emotionally competent, and innovative. The networked cosmopolitan prospective identities fabricated by the psychotechnologies of the curriculum of the future are not "natural" or nonpolitical categories. They have been made up through the expertise of psychological science and computer science as an ethical-economic-psychological mode of life to be organized in the future of school. The extent to which the normative ideals about a networked, cosmopolitan social, and political order—personified by the networked cosmopolitan identities promoted by Q2L and Learning Futures—constitute a desirable future for the school curriculum remains an open question. In reserving judgment at this stage, this chapter has aimed to give some indication of the complex webs of concepts, explanations, images, vocabularies, and theories that are increasingly shaping how the curriculum is thought, influenced, and acted upon, and to question those things as if they are natural and taken for granted.

PART II

The Politics of Technological Opportunity and Risk

CHAPTER 4

Becoming the Future: The "Lake Highlands" Middle School Laptop and Learning Initiative

David Shutkin

Introduction

In the fall of 2008, the "Lake Highlands" School District unveiled its "Laptop and Learning Initiative." Creating a 1:1 ratio of laptop computers for middle school teachers and students, the Lake Highlands' initiative was an expression of the district's commitment to the future. In their own words, this future requires the preparation of students to participate as well educated citizens in an increasingly interdependent and technologically advanced global society. This chapter is not about the future. Instead, it is about the "present of the future" when digital technologies are deployed in school districts to synchronize time and domesticate otherness. The synchronization of time is a technological response to the past, to determine the future by controlling the present (Simpson, 1995). Contrary to a synchronization of time, Lévinas (1987) describes time as a relationship between self and other, between a present that is experienced by and known to the self and a future that is unknowable and other.[1]

> The future is absolutely surprising. Anticipation of the future and projection of the future...are but the present of the future and not the authentic future; the future is what is not grasped, what befalls us and lays hold of us. The other is the future. (pp. 76–77)

This chapter is about this other; it is about middle school students in the Lake Highlands School District becoming the future as they experience the synchronization of time in school with their new MacBooks.

The Lake Highlands is an economically and racially diverse inner-ring suburban district located in a metropolitan area in the Midwest region of the United States. Seventy percent of the students in the district are African American; more than 50 percent qualify for federally funded free or reduced cost breakfast and/or lunch. During the 2008–2009 school year, less than 45 percent of the households of students in the district had access to the internet. And during this same 2008–2009 school year, the average level of proficiency in grades 6–8 across the 3 middle schools was 56.4 percent in mathematics and 63.5 percent in reading.[2]

Through narrative inquiry and actor network theory, I explore the mutual conditioning of information technologies, educational policy, and federal legislation in the conceptualization and implementation of the Lake Highlands' Laptop and Learning Initiative. Narrative inquiry is a naturalistic and self-reflective research methodology that emphasizes lived experience and local context (Conle, 2000; Clandinin & Huber, 2002). The primary strategy is to learn the stories and perspectives of the people involved through observation and participation. Toward these goals, I have been working with teachers, students, administrators, and technical support personnel in the district for about a year. And through this process, I am learning about and incorporating into my research, participant knowledge, skills, and perspectives about technology, education, and the historically specific context of the Lake Highlands School District. As a technologist too, I have searched for a means to better understand how different technologies are implicated or participate in the history or practices of everyday life. In this instance, my research questions and participant mode of inquiry have led me to engage in a narrative form of inquiry that brings me closer to actor network theory (Latour, 1991, 1993; Strathern, 1996; Brown & Capdevila, 1999; Law, 2008). It is from this perspective that I explicate how the attributes of certain technologies, in conjunction with federal and local policies, the myriad actions of school personnel, and other elements of the network constrain and condition this reform initiative.

Almost immediately following the rollout event when the students received their MacBooks, the focus of the Lake Highlands' laptop initiative shifts to a new temporal register, less concerned with preparing students with the "twenty-first-century skills" needed to succeed in the global economy. In this new register, the focus is on risk and the immediate future. The emphasis of the initiative also shifts abruptly from the new MacBooks to the district's internet filter. Used to block access to social media such as Facebook and YouTube, the internet filter is also used to block web 2.0 design, publishing, and networking applications that would otherwise afford opportunities to advance student achievement. In the concluding section, I discuss a social studies class where middle school students were assigned a digital media research project. Reflecting on this experience, I conclude that the Lake Highlands' Laptop and Learning Initiative is characterized simultaneously by a technological openness to the future and by a domestication of time (Simpson, 1995). This ambivalence includes a sociotechnical invitation to the students to be made subject to the laptop initiative and the district's vision of the future. At this juncture, I make a return to Lévinas and the present of the future to characterize this sociotechnical invitation as a reduction of the other to the same, as an aligning and conceptualizing of the future in terms of the present. To the extent to which these middle school students use anonymous proxy servers during class to bypass the internet filter and surf away from class to more interesting places, perhaps these students are refusing the district's vision of the future and, instead, are choosing to forge an alternative present and a future of their own?

In the Present: Preparation for the Laptop Initiative

A team of administrators from the Lake Highlands School District, led by Greg Loyda, Director of Educational Services and coordinator of the MacBook initiative, approached the chairperson of the department of education at my university to discuss the feasibility of enrolling district teachers as a cohort in a graduate-level course to advance their one-to-one laptop integration initiative. As a participant in these meetings and an educational technologist, I was invited to research and design an initial course to prepare Lake Highlands' middle school teachers and other district personnel to design and implement technologically enhanced learning experiences with laptop computers and to help them integrate the laptop initiative across the middle school curriculum. As part of my research, I met with middle school teachers and principals (prospective students for the course) to learn about their learning and teaching styles, their needs as educators and perspectives about technology, including their thoughts and assumptions about the MacBook initiative. With some precursory course design ideas sketched while reflecting on these experiences, I arranged a standing meeting with Greg Loyda, my contact person in the district.

During our summer meetings, two ideas came together including the district's embrace of twenty-first-century skills and the pedagogic potential of so-called web 2.0 applications. These skills were organized around the following general themes: creativity and innovation; critical thinking and problem solving; communication and collaboration; information, media, and technology; and life and career (Partnership for twenty-first century skills, 2009). Web 2.0 applications, including tools for blogging, networking, information organization, and so much more are web-based design, publishing, and collaboration applications. During our summer meetings, Mr. Loyda and I agreed that I should explore the intersection of twenty-first-century skills and a broad range of web 2.0 curricular and pedagogic technologies with the teachers and administrators that were to enroll in the course.

Rollout Night at the Middle School

With the course scheduled for the spring semester, that fall I was invited to attend the MacBook rollout event at Jefferson Middle School when every student was to receive a MacBook. It was a cool evening in October. After dinner, I rode the city bus several miles to the corner of Royal and Jefferson. There, on the west end of the Lake Highlands, I walked a few blocks to Jefferson Middle School. Once inside the building, I followed a series of signs down a narrow corridor leading to the main office. Institutional posters meant to inspire with pithy phrasing and bright colors were evenly spaced along drab walls. As I walked, I stopped to read a few of them such as *The Lake Highlands: Preparing All Students for Success in a Global Economy*. A few paces on, I stopped to read again, *Minds Opened, Talents Discovered, College Ready, Future Prepared*. I arrived at the office where I signed in and received a nametag to stick on my jacket. As I was walking out of the office, I was greeted by Greg Loyda. He asked me if I knew Virginia Neal, the superintendent of the Lake Highlands School District. Just as I was saying "no," he gestured for me to turn and introduced me, "I'd like you to meet Dr. David Shutkin from John Carroll University. He's designing a course for teachers participating in the laptop initiative." We shook hands and exchanged pleasantries. It was then that I learned that it was Dr. Neal's vision to establish the first one-to-one laptop initiative in an urban school district. And tonight she was the featured speaker.

Moments later, I entered the auditorium where the evening's events were scheduled to begin. Hundreds of middle schoolchildren were there with their parents. Until the lights were dimmed, there was a constant buzz of conversation; I could feel the children's excitement. To the cheers, smiles, and rousing enthusiasm of everyone in attendance, from parents, from their children, and from the many invited guests, Dr. Neal rose from her chair, walked to the podium, and welcomed everyone to the evening's event. In a brief speech, she offered her vision of the Lake Highlands' laptop initiative. She maintained throughout that the new responsibility of the district and its educators is to prepare the children of the Lake Highlands for the future:

> The twenty-first century is emerging as a century of global trade, where more children speak English in China than in The United States. It is emerging as a digital century and our children must be prepared with the skills to compete... To safeguard the future for our nation, our children need to apply their new technical skills to develop the critical thinking and problem solving skills to compete in the global market place... In this era, to educate our children, we cannot use yesterday's teaching methods to teach tomorrow's knowledge workers the skills to safeguard our nation's future; boredom in school must give way to engagement and to this end we need the Lake Highlands Laptop Initiative.

As she stepped away from the podium, the auditorium filled with applause. Students and their parents quickly filed out of the auditorium and dispersed to three separate rooms for the next event listed on the program. I followed a family walking ahead of me. We were being led by a bright-eyed African American girl, maybe 13 or 14 years old. She walked with purpose in her gait to a smaller gymnasium with cafeteria style tables in four long rows. Once in the gym, a digital image filled a floor-to-ceiling video screen with the school's mascot and name. Below the mascot were the words, "Safety First."

I sat across the table from the young girl and introduced myself to her and her mother. Her name was Latoya. She sat upright, her hands folded on the table. The quieting voice of Kevin McManus filled the room. An industrial arts teacher hired in the early 1970s, in his current position, Mr. McManus is one of three technology-curriculum integration specialists in the district. He outlined a series of internet "risks" and issues associated with the proper care and maintenance of Apple MacBooks. He browsed to several informational websites, showed a video, and in all outlined a broad range of issues and topics: the Children's Internet Protection Act (CIPA), internet pornography, online predators, Facebook, instant messaging, filtering software, restricted access, and so on. To the room filled with scores of excited students and their tired parents, he asked if anyone had any questions. No hands were raised.

Our attention was then directed to the back of the gymnasium where several large carts were stacked with boxes upon boxes of Apple MacBooks. Parents were instructed to open their folders and pull out several forms, including the acceptable use form. A pen was provided. Each student was instructed to bring the signed forms up to the carts when their name was called. A simple exchange and a check mark by the student's name would insure that everything was done according to plan.

The Synchronization of Time

After saying good night to Latoya and her mother, to Greg Loyda and a few others, I left the building and started on my way home. As I did, my mind wandered from thoughts of internet safety and risk management to Dr. Neal's speech. I thought about the movie *The Matrix*...no, it's the wrong metaphor, the students are neither to be future batteries nor are they prisoners in Plato's Cave.[3] Maybe *Star Wars, Attack of the Clones*? This comparison is a little closer; with an emphasis on technology, these middle school students are to be educated, prepared to serve the nation. They are in a state of becoming; but this isn't science fiction; these are not clones. No, this is neoliberalism. My mind jumps to Heidegger's (1977) treatise on technology. With Dr. Neal's words echoing in my mind, "safeguard the future...nation,...children...technical skills...global market place," I begin wondering about the students. Are they positioned as Heidegger's *standing reserve*?

This latter comparison is rather crude and essentializing by definition, and I do not necessarily expect it to hold up to scrutiny. Nevertheless, the significance of technology for Heidegger is associated with its use as a means to extend instrumental rationality, to find the most efficient or cost-effective means of achieving a given end. Heidegger describes instrumental rationality as a technological clearing; it is the shared practices and beliefs, or "mind-set," into which people are socialized. It is an understanding of humanity, of things and society that comprises a universal understanding of "Being" (Dreyfus, 1995). This technological clearing is the maximization of flexibility and efficiency for its own sake. The goal is the optimization of people, things, and society, "We pursue the growth or development of our potential simply for its own sake—it is our only goal" (Dreyfus, 1995, p. 101). Averse to any reflective tendency, instrumental rationality, according to Heidegger, informs the modern mind-set as it conceptualizes the world as "standing reserve"—from the natural gas trapped in the shale formations deep beneath our park lands to the students less than fully engaged within our classrooms.

Simpson (1995) relates Heidegger's technological clearing to the question of time helping me better understand Dr. Neal's emphasis on the future. The most basic assumption of this technological clearing, Simpson explains, is the linear, synchronization of time. It is the application of technology to the prediction and control of the future and the secure ordering of the present in response to a past that defies control (Simpson, 1995). The synchronization of time is an integral practice of Western thought through which time is made linear, is sequenced, and incorporates origins. The past merges into the present, which anticipates the future, and both future and past serve the understandings of the present. Certainly, Dr. Neal was not the first school administrator to conceptualize technology integration as a future-oriented school reform initiative. Indeed, the deployment of technology in the field of education is frequently conflated to notions of progress and the future (Shutkin, 1998). Following Simpson (1995), the modernist present is conceptualized in terms of past performance with an intention to forecast the future. In this way, Dr. Neal's laptop initiative, in the present tense, could be thought of as an endeavor to orchestrate a technology integration initiative in the district as a response to a past marked by persistent failures to realize adequate yearly progress, a measure of achievement established by the US Federal "No Child Left Behind" Act. This synchronization of time is instrumental to both the administration of schooling and to common understandings of things future and past, of computers and people, including middle school students.

Translating Time into Networks

In my efforts to design and teach my course on technology integration for teachers and administrators in the district, I spent much time as a participant observer in the middle schools across the district helping teachers and students integrate technology into their classroom learning experiences. Throughout, a theme of time and technology emerged as the goal, use laptop technology to prepare students with twenty-first-century skills that they will need in the future. Yet, as I would reflect on my experiences with teachers, principals, and students in the district, it seemed that Heidegger's ideas about the technological clearing and Simpson's efforts to relate this clearing to a modernist synchronization of time were too essentializing. No matter how hard everyone tries, no matter their integrity, school reform initiatives complicated by educational technology integration do not easily conform to linear models or essentializing philosophies of instrumental change. Following Feenberg (1999), for Heiddeger, modern technology is not a contingent or emergent development in history, rather it is both transhistorical and determinant of social relations. Further, according to Heiddeger, modernity is determined by this abstract form of technical action or instrumental rationality and no matter what happens (socially, politically, or historically), it remains unchanged. Feenberg (1999, p. 15) writes, "I call this 'essentialist' because it interprets a historically specific phenomenon in terms of a trans-historical conceptual construction." Additionally, in Heidegger's

(1977) discussion there appears to be a sleight of hand; the technological object seems to disappear or to somehow become subsumed within or as a part of instrumental thinking. Yet, levers, bridges, hammers, and computers do not refer to the same things or describe the same practices.

It was obvious to Dr. Neal, Greg Loyda, and the director of Information Technology, Dr. Keith Broadnax, that the MacBook technology was protean in nature (Koehler & Mishra, 2008). In other words, laptop technology is characterized by multiplicity; it is so many things unto itself and for its users as well. Middle school students use their laptops in ways that differ dramatically from their teachers, parents, or school administrators. Of course, the MacBook was used by teachers to manage their online grade books, deliver online textbook materials, prepare and give keynote presentations using video projectors or Smart Boards, assign drill and practice exercises, and so forth. Just as easily, the MacBook was used by students for taking notes in class, video chatting, surfing the web, and updating their Facebook pages. This type of student use was anticipated by Dr. Neal and by Dr. Broadnax and, before them, by the CIPA of 2000, which requires schools to integrate an internet filtering system as a condition for receiving federally funded E-Rate money.[4] Thus, for the Lake Highlands School District to prepare their students for the future they would need more than just MacBooks; they would need to orchestrate a technology integration initiative across three middle schools. Indeed, a multiplicity of people, technologies, policies, beliefs, and so on were being enrolled into the initiative forming an intricate web or network of heterogeneous and mutually conditioning elements (Foucault, 1985; Law, 2009).

The concept of the *network* used in actor network theory describes a very different sense of the temporal than Heidegger's linear synchronization of time. A linear and modernist synchronization of time deploys technology to forecast the future, to design the present as an accounting of an unwieldy past, and to best prepare for the future (Simpson, 1995). An actor network is genealogical (Foucault, 1988) not linear; it evolves as a heterogeneous ensemble of mutually conditioning and interconnected elements whose interaction effects, if there are any, are difficult to predict. Central to actor network theory is the assumption that to understand technology or technoculture is to assume that there is no "inside" or "outside." Technology is no more separated from the practices of industry or government than is "education." The discourses and practices of technological development or the diffusion of innovations are unbounded processes indiscriminately crossing disciplinary boundaries within and beyond the academy. Actor network theory refuses to distinguish the macro from the micro and instead thinks in terms of a network extending rhizomatically from the microscopic in some instances to the global, and frequently in the same instance. In the context of an ozone network, Latour (1993, p. 11) describes that the network would link "in one continuous chain the chemistry of the upper atmosphere, scientific and industrial strategies, the preoccupations of heads of state, the anxieties of ecologists."

Latour's descriptions of the ozone network are fixed in time; in this form, they are just a descriptive "snap shot." Conversely, networks are dynamic and refer to intentionality and action such as researching in a science lab, developing green technology or, as in this chapter, integrating educational technology. The signifier *network* in this way incorporates a temporal dimension referring to a series of changes through time and into the future, researching, developing, and integrating. Latour initially suggested the concept of *transformation*, which offers a certain temporal clarity but settled instead on the term "translation" to imply the intentionality of the temporal emergence and evolution through time of an actor network.

Drawing from the field of semiotics for explanatory power, the temporal process of translation combines two moments, the paradigmatic and the syntagmatic (Barthes, 1968; Fiske, 1990). Following Saussure, Fiske (1990) defines a paradigm as a set of signs from which only one can be chosen to be used at a given time. For instance, from the set of all possible laptops, the Lake Highlands chose MacBooks for their one-to-one initiative. A syntagm is the message or meaningful grouping into which a plethora of signs are organized. This grouping can be linguistic, material, or both. In this way the linguistic syntagm, "The Lake Highlands School District chose MacBooks," is easily distinguished from the linguistic syntagm, "The Shoreland School District selected Asus Laptops." Syntagmatically, the material process of translation, within actor network theory, describes the formation of the network through time as the enrollment of elements (not limited to people, policies, and technologies) into a project with purpose and intentionality. Latour and his colleagues draw on the work of the semiotician A. J. Greimas and his "semiotics of the natural world" to describe the material process of the translation of actor networks (Høstaker, 2005; Law, 2009).

Internet Filters, Risks, and the Agency of Things

Technology detaches us from the past that is encoded in our present and draws us ever more into a preoccupation with what lies just beyond our world in the future. (Davison, 2001, pp. 154–155)

The future, in its relation to technology, is multiple, shifting, dynamic. Indeed, as I establish in this section, almost immediately following the rollout event, the present tense of the Lake Highlands' laptop initiative that emerges through this sociotechnical translation shifts to a new temporal register, to a different experience of time, less concerned with preparing students with those new skills needed to succeed in the global economy (Shutkin, 2005). Instead, in this new register an issue assumed by district administrators to have been managed and under control suddenly reemerges to refocus attention on risks and the immediate future, on tomorrow or next week or next month. Emphasis shifts abruptly from the MacBooks and the rollout event to the internet filter, first to safeguard the students' online experiences from internet predation and the dangers of pornography. And, as I emphasize in this section, the internet filter's use is extended to block access not only to social media including Facebook and YouTube but also including web 2.0 design, publishing, and networking applications that afford so much promise to classroom practice.

In the educational technology course I was designing as a complement to the district's laptop initiative, web 2.0 applications were to be a pedagogic feature and a central part of the curriculum. At the time, web 2.0 was new and exciting for me, for Greg Loyda, and for the teachers, principals, and curriculum specialists enrolled in the course. During the summer before the rollout event, Greg Loyda, Director of the Lake Highlands Laptop and Learning Initiative, and I shared numerous productive conversations about the role of web 2.0 applications in education. Across the country and around the world, teachers were designing educationally meaningful and pedagogically significant strategies for integrating Facebook, YouTube, PBWiki, VuVox, and many other web 2.0 applications into classroom practice (Prensky, 2008; Jenkins, Purushotma, Weigel, Clinton & Robinson, 2009). Yet many other educators emphasized the risks these applications present to the integrity and security of school networks, to the realization of educational goals conventionally conceived, and to the safety of students (Goldmann, 2007; Wolinsky, 2008; Millea, Galatis & McAllister, 2009).

The internet has also made pornography perniciously invasive even into the lived experiences of those most vulnerable and innocent (USDE, 2010). Politicians at the federal level therefore identified a need for legislation to protect children attending schools and in the year 2000, after several failed initiatives, the US Congress passed the CIPA. As federal law, CIPA imposes regulations on any elementary or secondary school that receives funding for internet access through the federal E-Rate program. To be in compliance with CIPA, schools must establish an internet safety policy that includes "technology protection measures" to block or filter access by minors to content that is deemed to be harmful to them. According to the law, the determination of what material is harmful to minors is to be made locally by whatever authority in a school district is responsible for making the required determination. It is stated explicitly that this determination is not to be made at the federal level (CIPA of 2000).[5] The technology protection measures used by the Lake Highlands School District took the form of an internet filter developed by S62 Protective Services, Inc.[6] Conventionally, the filter is used to protect students from web-based content determined to present risks to their well-being. The use of the S62 Internet Filter is in accordance with CIPA as well as the district's acceptable use policy. Under the auspices of the Lake Highlands Board of Education, the director of Information Technology, Dr. Keith Broadnax, an attorney for the district, manages and coordinates the integration and use of the S62 Internet Filter. Dr. Broadnax is responsible also for designing, deploying, operating, and maintaining the information technology infrastructure throughout the school district. This includes acquiring the federally supported E-Rate funding for accessing the internet, maintaining the district's new wireless network, and managing a qualified but typically understaffed technical crew. As an attorney, he is also required to interpret and respond to the CIPA, which includes formulating and implementing the district's acceptable use policy and integrating the S62 Internet Filter into the network.

During our conversations in the summer before the rollout, Dr. Broadnax was emphatic about issues of risk management and the avoidance of any legal action in relation to the slightest probability that any child might stumble upon a website with "inappropriate" materials. More explicitly, as defined by the CIPA, inappropriate materials refer to "visual depictions" of obscene material, child pornography, and materials that are harmful to minors. The S62 Internet Filter, as a technology protection measure, is enrolled by the school board and district administrators into the Lake Highlands' actor network that also includes the MacBooks, district-level technology plans, technology service contracts, internet safety policies, and the 2001 CIPA. Translated through time from the 1990s to the early 2000s to the present year, 2013, ideas, beliefs, fears, and risks that relate technology to education, to a sense of social justice, and to morality are made durable, first in the form of federal legislation, and then as technologies—routers, internet filters, and so on (Latour, 1991; Strathern, 1996).

In this context, it is significant to recognize how technology relates variously to time. Conventionally, the one-to-one laptop initiative invites teachers, students, and their parents to conceptualize technology in relation to the future, as everything new, inventive, and forward. Yet in the year 2013, engineered into the design and functionality of the S62 Internet Filter are elements that trace historically from the 1990s; as a "protection measure," the S62 is a fold in time translating beliefs, risks, policies, and laws from the past into a technological element of the current Lake Highlands' actor network (Brown & Capdevila, 1999). The S62 protective measure is similarly the translation of risk of internet predation or pornography into technological practices enrolled in the Lake Highlands' actor network (Van Loon, 2002). While CIPA requires an internet filter, in practice the filter becomes a surrogate for this law. Reaching beyond the CIPA, the affordances and constraints of the S62, now enrolled in the Lake Highlands' actor network, harbor the potential for a broad range of "protective measures." Technically, the S62 is installed on a proxy server by the district's internet service provider. Requests to access websites from the internet must travel through this proxy server and the S62 Internet Filter before reaching the internet. The S62 filters the requests and returns only websites that are not blocked. Further, the S62 Internet Filter does not serve the district in a purely automated or neutral fashion. The S62, like most filters, is not completely automated; it requires much human labor to work effectively. Indeed, as each new website is identified by employees of S62 Protective Services, Inc., it is reviewed and categorized according to its form and content. In this way, the filter is updated every night.

Configured for the Lake Highlands School District by Dr. Broadnax, the S62 Internet Filter organizes websites into three main categories. First, there are CIPA websites that are blocked by all school districts because they are obviously inappropriate or harmful. Next are the Lake Highlands' websites available on the district computer network for use by teachers and students. The third category includes the Lake Highlands' websites that are blocked to maintain the "integrity and security" of the district network. As Dr. Broadnax explained to me, a website is placed in this third category at his discretion, if he determines that it poses a risk to the integrity and/or security of the computer network. Even as the S62 is enrolled in the Lake Highlands' laptop initiative as a surrogate for the CIPA, in its hybrid alliance with Dr. Broadnax, the S62 takes on broader responsibilities (Brown & Capdevila, 1999).

Interestingly, in this instance, Dr. Broadnax chose intentionally to use the S62 Internet Filter to block access by students as well as teachers to specific websites. He asserted the district policy to strictly control information published to the internet by students, teachers, or principals that risk association with the Lake Highlands School District. In this way, Dr. Broadnax configured the S62 to block all web 2.0 applications. This included site blocking of web 2.0 applications with demonstrated pedagogic potential and relevance such as Google Docs, PBWiki, PollEveryWhere, and Ning (Millea et al., 2009). As I was learning about this, the technology integration course that I had designed for the district was well underway. In an email communication to teachers and principals enrolled in the course, Dr. Broadnax explained that the risks to the school district are too great to use these types of web publishing tools.

In the email, discussed in class that week, Dr. Broadnax explained that some of the content his students published to the internet appears to contain information that competes with current Lake Highlands School District websites or information not in alignment with district policies. He then outlined the district's policy to restrict web-based information to servers owned and maintained by the school board or by an approved third-party host. Without these controls, he claimed, personally identifiable student information was at risk. He explained that search engines could index these personal sites leading people to believe that they are the Lake Highland School District's official websites. Dr. Broadnax asserted further that these personal sites were vulnerable and could be "hijacked" as a way to gain access to more people. He then insisted that everyone enrolled in my course remove his or her content from any third-party website at the completion of each assignment. As a gesture of goodwill, he did offer them space on servers maintained by the district for hosting individual teachers' web content and added that personnel from the IT department or a technology curriculum integration specialist could assist them in publishing their content on district servers.

While the Lake Highlands School District is appropriately vigilant in safeguarding students from the preponderance of inappropriate web-based materials, is in compliance with CIPA, and has an acceptable use policy, I learned from Lester Brafman, president of the district's teachers union, that there is neither a district policy nor a set of procedures governing the interpretation of CIPA including the deployment of an internet filter. Instead, these decisions are made exclusively by Dr. Broadnax, a common practice across many school districts (Wolinsky, 2008; Millea et al., 2009).

Dr. Broadnax's email communication functioned to undermine the will and determination of the teachers, principals, and curriculum specialists in my course who had worked for many long hours to develop both a comfort level with numerous and varied web 2.0 applications and the ability to produce technically enhanced learning environments with web 2.0 applications designed to advance the learning of their students. Of course, it was gracious of Dr. Broadnax to offer to publish their work on district servers. However, due to his emphasis on mitigating risk, what he could not understand about web 2.0 was the ubiquitous, interactive, and engaging potential that was lost to these educators and to their students in the Lake Highlands. As described in a report published by the Australian government, "Web 2.0

site blocking in schools is a risk management response to the difficult and not well understood issues that schools face in trying to balance cyber-safety concerns with the desire to harness innovative web 2.0 style collaborative teaching and learning" (Millea et al., 2009, p. 2). Later the report suggests some of the effects that this site blocking is having on the efforts of teachers to engage students, to advance their learning, and to explore multimodal literacies through the design of technologically enhanced learning environments:

> Despite the desire of some teachers to explore the benefits of web 2.0 for creativity and learning they are constrained by educational authorities and restrictive school policies framed around risk-averse practices in fear of legal action. Overall literature reveals that society seeks to protect children from perceived dangers of new media at the expense of creative, innovative and collaborative learning through the use of tools available in the twenty-first century. (Millea et al., 2009, p. 15)

Risk is not a person, it is not technology, and it is not even a tangible object. Yet risk, as the possibility of experiencing real danger or suffering harm, has always already been enrolled as an influential element in the Lake Highlands' laptop initiative (Van Loon, 2002). This risk emerges in a unique relationship with Dr. Broadnax. Together they form a "hybrid alliance"—that is, an alliance between a thing and a human subject (Brown & Capdevila, 1999). A thing such as risk (or the S62 filter) is devoid of subjective intentions or desires. Yet risk seems to derive agency through its capacity to inform the actions of others. While this seems self-evident, in the absence of intentionality, the thing provokes in a human subject a will to connect, to make use of, to relate to, and to form an alliance with the thing. This relationship forms, as Brown and Capdevila (1999, p. 4) write,

> not on the basis of what is said or done, but on the basis of what might be possible what may be subsequently presented. For are not all alliances at base promises about some future event? Do not all relations begin with an orientation toward what may be granted to the partner at some unspecified point(s) to follow?

This thing called risk is the prospect of realizing danger or suffering harm but then this risk, as an element in the Lake Highlands' actor network, can never take action. As Van Loon (2002, p. 29) observes, "Risks are always threatening to take place, they never take place (as disasters do). They are events in becoming." In this instance, the agency of things, of risk or an internet filter, is based on the formation of a hybrid relationship and the potential to take action at some time in the future.

In relation to the Lake Highlands' laptop initiative, risk emerges in two primary forms. Recalling the words spoken by the superintendent, Dr. Neal, during the rollout night, the laptop initiative is motivated by risks to the future economic development of the United States. However, there is a marked shift in register from this instance of risk, which places emphasis on a distant future relative to the present experience of students in schooling, to a more immanent risk threatening the secure functioning of the Lake Highlands' laptop initiative. The pedagogic and curricular implications too are dramatic for the laptop initiative, which was to feature the integration of the MacBooks together with web 2.0 applications.

The Domestication of Time, Ambivalence, and the Irony of the Present: Revisiting Rollout Night

Dr. Neal's emphasis in her speech during the rollout night on technology and the future conforms to Heidegger's descriptions of a linear, synchronization of time; technology is applied in the present to predict and control the future in response to an unruly past (Simpson, 1995). This is evidenced by the twin emphases, during rollout night, on the "open" potential of the MacBooks and on planning for the risks that these laptops present to the students and to the district. Simpson (1995) discusses this temporal ambivalence as he writes,

> Technology, through its incorporation of the temporality of planning… is essentially oriented toward the future, a future which is understood as an open horizon. However, this is an openness about which the technological project is deeply ambivalent. This understanding of the future is compromised by technology's rancor against the uncontrolled past and its concern with predicting and controlling the future. The goal of technology is the "domestication" of time, that is, the prediction and control of that which appears in time. Domestication appears as the will to control. (p. 53)

In the gymnasium following Dr. Neal's speech during the rollout night, I witnessed a foreshadowing of this theme of ambivalence that has come to inform this initial year of the laptop initiative. When Kevin McManus, the technology integration specialist, had finished outlining the district's concerns about internet risks and the proper maintenance of the MacBooks, he directed our attention to the back of the room. There we saw several large carts stacked with boxes upon boxes of Apple MacBooks. Parents were instructed to open their folders and to first read and then sign several forms, including a copy of the district's acceptable use policy.

When Latoya (the student I first introduced in the rollout section) heard her name, a prominent smile appeared on her face. Before she stood up, she turned her gaze to her mother seeking approval. Her mother nodded with confidence. With the several forms signed and neatly in hand, Latoya walked to the back of the room to receive her MacBook. For Latoya, like so many of the several hundred middle school students in the Lake Highlands School District, over the course of that first year of the laptop initiative, the meaning of their MacBooks was constrained by the many actions of their teachers and school administrators, and in relation to the enrollment of the S62 Internet Filter, the CIPA, and risk. More than that, the MacBooks became *thematized* through their enrollment in the laptop initiative. Strhan (2007), referencing Lévinas, explains that to thematize implies the act of offering to the other. What is offered to the other becomes imbued with so much meaning, not unlike the sociocultural meanings ascribed to gift giving (Bourdieu, 1990).

Throughout this initial year, the thematizing of the MacBooks, characterized simultaneously by a technological openness to the future and by the domestication of time, includes an invitation to the students to be made subject to the laptop initiative (Strhan, 2007). At issue is whether or to what extent these middle school students have internalized (accepted the invitation to) the vision of the future espoused by the district and made manifest through their experiences of the initiative. Lévinas describes thematizing practices as totalizing initiatives that reduce the other to the same. He writes, "Thematization and conceptualization, which moreover are inseparable, are not peace within the other but suppression or possession of the other" (as quoted in Strhan, 2007, p. 413). Lévinas (1998) emphasizes that within Western thought, the relation of the knowing subject to the other is an assimilation or domestication of the other. Yet, could it be that these middle school students are choosing to interrupt this invitation and thus forging an alternative present in preparation for a future of their own making?

Metamorf and the Irony of the Present

Once in the hands of the middle school students, it became exceedingly difficult to "domesticate" the MacBooks or the students. Indeed, as I observed toward the end of the school year, while the district might have deployed the S62 Internet Filter to block student access to inappropriate websites and required them to sign the acceptable use policy, together with their MacBooks, the students were more likely to ignore the policy and to play the S62 Internet Filter as a rather simple level of a mundane video game (Gee, 2008). With their knowledge and wherewithal to use anonymous proxy servers, the S62 was rendered ineffective by the students.[7]

> You are going on vacation! All students have a city of their choice somewhere in the United States of America. You will travel from The Lake Highlands to your city. The guidelines listed below are provided for you to maximize your trip [sic].

Thus begins the description of "Travel Project City," a technologically enhanced learning experience designed by David Hecht for his eighth-grade social studies class at Ripley Middle School. David invited me to discuss the laptop integration initiative and to observe in his classroom as he and his students worked on Travel Project City. Though Mr. Hecht was present to answer questions and to support student learning, the entire time that I observed was dedicated to independent work on the Travel Project City assignment. While I was there, I walked around the classroom to observe and to assist the students; perhaps I could answer a technical or a conceptual question? There were some students diligently on task and fully engaged in the Travel Project City assignment. I sat with two boys who were really struggling with the project. I helped them read the assignment and to search for pictures of Dallas, Texas (the city one of the boys had been assigned to research) and to then insert the pictures into a KeyNote slide. Most of the students, however, were not engaged with the assignment. On more than one occasion, I counted upward of 13 students out of 22 wearing ear buds; I assumed they were listening to music or otherwise disengaged from the project. I observed students playing online video games, some were chatting and updating their Facebook pages, some were watching cartoons, and others were streaming World Federation Wrestling into their browsers. Almost everyone was texting or chatting online. It was more than evident to me that many of these eighth-grade students had well developed technical skills including the skill to use anonymous proxy servers to circumvent the S62 Internet Filter designed to prevent them from surfing the internet at will.

The project description handout had so many typographical errors; clearly, Mr. Hecht was not concerned with grammar, syntax, or usage. Neither were the students. However, he did require the students to insert music in their projects, "You must have 1 songs [sic] that plays at some point during your presentation." Many students, of course, were interested in this part of the project. While there were no expectations that the music would relate thematically to the city the students were researching, the students applied their own sophisticated cultural aesthetics to determining which song to insert. As I continued walking around the class, I observed a familiar student working very intensely. It was Latoya. I pulled up a chair to learn what she was doing. She said she had found a great song for her project but that she did not want to just link to or embed the YouTube video. Nor did she want to spend her own money downloading the song from iTunes. Instead, she was inclined to use a "free" (and apparently illegal) online web 2.0 application called Metamorf, to "rip" the song from YouTube.[8] As she worked, I quickly observed that this was a complicated process with multiple steps. Latoya was applying a sophisticated set of technical skills. No one else was doing this. When I asked her how she learned to do this, she said, "Mr. Hecht taught me."

Later, when I was talking with David he said, "Metamorf is great and it's not blocked from the network because, well, the IT department doesn't know about it. It's great for ripping MP3s from YouTube videos." He shared that the principal had shown it to him. The principal, Mary Collins, had learned how to use it when she wanted to use a song for a presentation to the school that had the word "ass" in the lyrics. Another teacher had "guided" the principal to Metamorf. He taught her how to use it to first rip the song and then to "clean up" the language. Later, the principal remarked to me rather flippantly, "Oh yeah, Metamorf is really cool."

Conclusions

Lévinas's earliest phenomenological explorations of the other emerge in his first writings about time. For Lévinas (1987), the future is absolutely other because it cannot be known, described, or experienced in the present tense. Lévinas discusses the mystery of the future as infinitely unknowable and absolutely other: "The future is what is in no way grasped...the future is absolutely surprising...The other is the future" (pp. 76–77). Then, as if stating the obvious, he concludes about time that it cannot be experienced alone; it is the experience of the self in the present with the other, with the unknown future, self and other, present and future. He writes, "The very relationship with the other is the relationship with the future. It seems to me," he continues, "impossible to speak of time in a subject alone, or to speak of a purely personal duration" (p. 77). The present is knowable to the self as the self; the future that transcends knowing is the other, and time is a relationship between self and other, between present and future.

Following Lévinas (1987), the thematization of the Lake Highlands' laptop initiative, through the discourses and practices of its actor network, through the relationships it sustains between self and other, between teachers and middle school students, is productive of the future. Yet, the Lake Highlands' laptop initiative cannot prepare these middle school students for an unknown, detached future (a reduction of the other to the self) any more than district administrators can determine the future by forcing students to conform to the district's vision of the future. Instead, through the middle school students' myriad experiences of the present in school, with their teachers and school administrators and with the varying elements of the Lake Highlands' actor network, the students are producing themselves as subjects of the future; they are the future.

I conclude that while the experience of schooling is to educate our students, present reform initiatives cannot be controlled with technology any more than the future can be predicted. The synchronization of people, policies, and technologies in a linear time gives way to the enrollment of elements in an actor network, to ideas translated through time into policies and policies translated into technology. In this way, the initiative to integrate web 2.0 technologies to advance twenty-first-century skills, as a primary goal of the Lake Highlands' laptop initiative, is undermined by the very elements most integral to the laptop initiative, namely the MacBooks, the internet, and the students. What proves to be most salient is the mutual conditioning of these elements with the district's internet filter, the CIPA, and the emergence of risk associated with access to the internet. Extending beyond Heidegger's (1977) thesis, Lévinas (1987) insists that time, as understood in Western thought, cannot be reduced to sociotechnical practices facilitating the prediction and control of the future. Rather, he asserts that the realization of time must also include an intersubjective relationship between self and other; time is a historical relationship between and among people, between a future/other that cannot be known and a self who is present.

Notes

1. This chapter was inspired by the phenomenological ethics of Emmanuel Lévinas. In particular, this chapter is informed by his book *Time and the Other* originally written while Lévinas was interned in a Vichy work camp in German occupied France during World War II. *Time and the Other* was first published in 1947.
2. Please note that in this chapter names of local people and geographic locations (and technologies where indicated) are all pseudonyms.
3. The film *The Matrix* is often compared to Plato's Allegory of the Cave. For instance, see Northwest Center for Philosophy for Children (2008).
4. Serving to shift the conditions and definitions of the digital divide, a provision of the 1996 Telecommunications Act continues to benefit school districts across

the United States by providing access to the internet at the discounted governmental E-Rate. To receive the E-Rate and associated technology and services,
school districts to this day are required to establish an acceptable use policy and
write a technology plan for the use and integration of digital technology.

5. While this is consistent with the history of states' rights and local governance
of education in the United States, it does seem to contradict the current and
largely bipartisan initiative to establish a national curriculum.

6. The S62 is a pseudonym.

7. Technically, an anonymous proxy server works as a transfer point between the
students' MacBooks and the sites on the internet that the students want to visit.
The anonymous proxy server hides the IP address of their MacBooks and even
provides encryption. Anonymous proxy servers were readily available on the
internet for the students to use.

8. "Metamorf" is a pseudonym.

CHAPTER 5

The Politics of Online Risk and the Discursive Construction of School "E-Safety"

Andrew Hope

Introduction

Following the introduction of widespread internet access into educational institutions across most—if not all—industrialized countries, it is possible to discern the common emergence of a school "e-safety" agenda, driven primarily by the publications of government-funded bodies and commercial organizations. In the United Kingdom—as in many other cases—central to this agenda is the notion that children are at risk online. In this context, "risk refers to the probability of damage, injury, illness, death or other misfortune associated with hazard. Hazards are generally defined to mean a threat to people and what they value" (Furedi, 2006, p. 25). Yet as this chapter will endeavor to show, these risks are not "absolute truths," rather they are selected (Douglas & Wildavsky, 1982, p. 29). This is not to deny that real dangers exist but rather to emphasize that a choice is made regarding which hazards to focus upon. Consequently, decisions regarding which risks should be causes for public concern, and the precise nature of such "hazards," are always political in nature. Thus, it follows that the school e-safety agenda should not be seen merely as an attempt to care for the well-being of children, but also to impose particular ways of seeing and behaving.

Central to this chapter is the observation that despite a primary concern with protecting children from online risks, they are largely excluded from the policy process—present as objects of study but rarely subjects of communication. This has resulted in the emergence at a national level of school internet-risk discourses dominated by adult perspectives. With this in mind, the chapter draws attention to the extent to which children's voices and viewpoints are peripheral in much of the e-safety agenda. Examining policy documents and commercial e-safety materials, it is asserted that the agenda around e-safety and internet risks has been to date a largely adult-centric exercise, which justifies coercive classroom practice as well as seeks to exploit new commercial opportunities. Ultimately, it is suggested that risk should be seen as property. The implications of such a change in perspective would be that children are no longer treated merely as policy triggers and denied the rights to participation. Rather their voices should be heard and their views considered. Not only might this change the very nature of the school e-safety agenda, but it would also mean that children's own concerns could be addressed as adults seek to hermeneutically understand children's actions and engagements with digital technologies.

The Byron Review and the Nature of Risk

The Byron Review (2008) provides a useful initial framework within which to critically explore some of the key elements surrounding the discussion of e-safety and the online risks facing children. Commissioned by the then UK prime minister Gordon Brown in September 2007, the Byron Review was an independent appraisal of the risks facing children from the internet and video games. The review, which was led by a high-profile consultant clini-cal psychologist Dr. Tanya Byron, published its findings in 2008, with an update of progress following in 2010. While noting that some risk categories may overlap and that boundaries are sometimes blurred, the Byron Review (2008, p. 16) utilized the risk framework developed through the "EU Kids Online I" (2006–2009) meta-analysis project (see table 5.1).

This framework is broad ranging in content and hints at a degree of social complexity, suggesting that children online can be recipients, par-ticipants, and instigators. Yet the framework was the product of meta-analysis of preexisting research that overwhelmingly privileged adult perspectives. It primarily represents an adult view of the problems facing children online.

The Byron Review had a notable impact on e-safety policy in UK schools as well as in wider society. It acted as a spur to a wide range of stakeholders, motivating them to become involved in a multifaceted, coordinated effort to address the internet risks that children were seen as facing. Yet, the review is also important because of the manner in which risk is constructed within its

Table 5.1   A classification of online risks facing children (Hasenbrink, Livingstone, Haddon, Kirwil & Ponte, 2007)

	Sexuality	*Values/ideology*	*Aggression*	*Commercial interests*
Content (child as recipient)	Pornographic or unwelcome sexual content	Bias; racism; or misleading information or advice	Violent/ hateful content	Adverts; spam; sponsorship; or personal information
Contact (child as participant)	Meeting strangers or being groomed	Self-harm or unwelcome persuasions	Being bullied, harassed, or stalked	Tracking or harvesting personal information
Conduct (child as actor)	Creating and uploading inappropriate material	Providing misleading information/ advice	Bullying or harassing another	Illegal downloading; hacking; gambling; financial scams; or terrorism

pages and the influence that this conceptualization might exert both nationally and internationally. Thus it can be discerned that the Byron Review (2008, 2010) utilizes the notion of "risk" in a particular manner, recognizing its subjective, contested nature, arguing that "risk-averse cultures" may increase children's vulnerability, suggesting that risk taking forms a normal part of the maturing process, and recognizing, albeit in a limited manner, that children can be a source of "danger" online.

Rather than adopting a realist risk paradigm within which the nature of the concept of risk is taken for granted (Lupton, 1999, p. 18), the Byron Review (2008, p. 17) notes that "deciding what is inappropriate is subjective and based on many factors including age, experience, values, belief systems and culture of the person making that decision." What one individual finds offensive might be seen as an important, empowering experience for another. Thus, the labeling of an online activity as "risky" is partly an issue of perspective, with disagreements arising concerning what constitutes a risk and what the possible impact on well-being might be. While broad social consensus might exist on certain risks, given the potential for conflicting interpretations, risk needs to be understood in the context in which it arises. Consequently, it would be naive to assume that the online risks facing children outlined in table 5.1 are those that will necessarily cause anxiety in a particular school. After all, individuals within educational institutions that cater for different age groups, sexes, socioeconomic groups, ethnicities, or religions may have distinct perceptions of risk, reflecting their own situated experiences and concerns. Furthermore, it does not follow that children will necessarily concur with adult interpretations of online risks.

Accepting that risk perceptions may be highly subjective and that differing viewpoints might flourish, the question of who gets to decide what becomes classified as a risk is of particular importance. Whoever can impose their risk perspectives on others, particularly if such anxieties become perceived as naturally the ones society should focus on, is in a position of power. The UK Office for Standards in Education, Children's Services and Skills (Ofsted, 2010, p. 4) notes that, having assessed e-safety in 35 UK schools, "few of the schools visited made good use of the views of pupils and parents to develop their e-safety provision." Thus, it would appear that at an institutional level, students and parents continue to have little influence upon what becomes regarded as an online risk.

It might seem strange to suggest that banishing online risks is not necessarily a desirable goal. Yet, risk-averse culture may actually increase children's vulnerability, denying them the chance to develop skills for identifying and coping with risks. Ofsted (2010, p. 5) found that "pupils in the schools that had 'managed' systems had better knowledge and understanding of how to stay safe than those in schools with 'locked down' systems." Furthermore, risk taking may have positive outcomes, forming a normal part of the maturing process and an essential element of identity construction. Indeed, Lightfoot (1997) maintains that risk taking has play-like qualities, offering children various opportunities, such as the development of social identity, fostering group membership, demonstrating autonomy, and adjusting to the requirements of adult life. In this context, risk should not be seen simply as a synonym for danger. Rather risk can be positive—a normal part of a healthy process of child development. Potential risks are closely associated with possible benefits online, as children experiment and push boundaries.

Although somewhat muted, the Byron Review (2008, p. 53) does acknowledge that situations exist where children are the instigators of inappropriate online behavior (the "perpetrator"). Such a distinction allows for a consideration that children online may not merely be at risk, but may also themselves be a source of danger. Indeed Staksrud (2009, p. 147) argues there is a growing realization that children can be a source of risk.

> The idea that children can be active participants in a negative sense through illegal or deviant [online] behaviour has received little attention...although issues such as "digital bullying," "happy slapping," and the illegal downloading of music and movies are starting to creep into the public—and official—consciousness.

This is not to be dismissive of online threats to the well-being of children, but rather to indicate that in certain situations, young people may be proactively and consciously engaging in behavior that could be labeled as risky by their peers as well as adults. Constructing children solely as "victims," merely being subjected to internet risks, is a somewhat limited analytical approach.

This brief overview serves to foreshadow some key concerns, such as how risk is defined, who is perceived as being "at risk," and whether risk is merely a synonym for danger. Somewhat lacking from this conceptualization are the notions that risk operates as a controlling discourse, legitimizes increased disciplinary technologies, is a commodity exploited by private security companies, and can engender resistance. While both the initial Byron Review and the subsequent update report include children's words and drawings, it is difficult to escape the conclusion that it still primarily represents an adult viewpoint of the online risks facing children. These issues will be explored as the focus turns to the broader e-safety discourse constructed through policy documents and commercial e-safety material. First, it is necessary to consider student internet-risk perspectives and the problem of "voice."

Schoolchildren, Online Risks, and Voice

Treating children as a generic group may represent a lack of subtlety in terms of socioeconomic, ethnic, or religious differences, but more strikingly it offers the possibility that the views of different age groups of children become confused or conflated. Consequently, a degree of critical wariness is required in interpreting any research or policy documentation that treats children's risk perspectives as homogenous and is primarily defined by their legal status as children. For example, examining the views of primary schoolchildren in the United Kingdom, Cranmer, Selwyn, and Potter (2009) suggest that the actual incidents that children label as risky tend to be "mundane" from an adult perspective, including concerns such as deleting documents, losing their place in games, getting bored, and angering their parents through mishandling error messages. Conversely, Cranmer et al. (2009) also note that when discussing potential risks, a tendency might exist for some primary schoolchildren to imaginatively exaggerate internet dangers, expressing concerns about hidden webcams spying on them, their house being burgled as a consequence of disclosure of information online, and sterilization arising from computer use. Thus, internet-risk perspectives for this age group could be described as ranging from mundane matters to the sensationalistic and fantastic. Yet care is needed to avoid merely labeling such views as inconsequential or exaggerated. Rather they may offer real insights into issues that concern students.

Thus from a practical point of view, the fears expressed by young people above may be indicative of broader anxieties about surveillance, blame for material loss, and health issues. Significantly, as Douglas (1966) infers, risk and danger also operate on a symbolic level. Thus, discussing the Lele people of what was then the Belgian Congo, she relates how despite death commonly arising from carnivorous predators, starvation, and disease, the Lele were most afraid of lightning. This should not be dismissed as an irrational view, but rather as providing insight into the values and beliefs embedded in the Lele culture. Similarly, the risk narratives presented by children need to be hermeneutically understood rather than dismissed as irrational if they do not match the adult worldview.

Hope (2011) illustrated that while students in UK secondary schools echoed staff internet-risk discourses to a limited extent, occasionally touching upon the problematic nature of pornographic material, chat lines, and school network security, the main internet-related issue that worried them was that they would be punished for "inappropriate online behaviour," possibly regardless of their actual conduct. This reflects the findings of Selwyn, Potter, and Cranmer (2010, p. 123) that primary school students' accounts of risk rarely corresponded with official notions of e-safety. Indeed in some instances, postprimary students engage actively in risky online behavior in schools to engender excitement, foster a reputation, and build group identity (Hope, 2007).

Children's experiences and views regarding the internet are very different from adults and despite this gap in knowledge it is adults who, often with little if any consultation, "make up the rules and control the access" (O'Neill & Hagen, 2009, p. 235). Yet simply privileging "student voice," with regard to school e-safety, can foster problems. Students' views might be co-opted to meet "new mangerialist" outcome targets, rather than engendering democratic agency. Thus Fielding and Prieto (2002, p. 20) argue that it is "crucial for students' perceptions and recommendations to be responded to, not merely treated as minor footnotes in an altered adult text." Furthermore, student voices might be unduly influenced by hegemonic, adultcentric discourses, raising further questions of authenticity. Also, in terms of inclusion it is necessary to consider which students' voices are included/excluded, who speaks for whom, the breadth of discussion, and the nature of the language codes used. Underpinning such considerations must be an awareness of power relations, particularly given that the degree of student inclusion, the nature of participation, and the role played by adults in relation to students will all be decided by adults. Ultimately such practices should not come at the expense of teacher's voices, yet the cost of disprivileging student views, both in terms of democratic participation and learning, should not be underestimated. After all, in practical terms strict limitations placed on students' online activities in school means that those without plentiful internet resources elsewhere might fail to develop "more sophisticated information-seeking skills" (Robinson, 2009, p. 505). With such considerations in mind, the focus will now turn to the e-safety discourse as constructed through policy documents and commercial e-safety material.

E-Safety Discourse

The following discussion draws upon Foucauldian discourse analysis, focusing on the power relationships as expressed in national school related e-safety policy documents and guidelines. While much of the analysis draws upon examples from the now defunct British Educational Communications and Technology Agency (Becta), which operated as the lead agency for the development and delivery of the UK government's Information Communication Technology (ICT) and education strategy for over a decade, data will also be drawn from the United States and European context. The subsequent discussion will focus on five emergent themes, namely (i) how risk is defined; (ii) the social construction of children in the documents; (iii) children's voices; (iv) the e-safety discourse as social control; and (v) the commodification of risk.

Defining Risk

In child e-safety documentation, risk is often utilized merely to denote danger or the probability of negative outcomes. For example, in *E-safety: Developing Whole-School Policies to Support Effective Practice* (Becta, 2005), risk is used simplistically in a realist manner as a synonym for danger, with issues such as copyright infringement, obsessive use of the internet, exposure to inappropriate material, inappropriate or illegal behavior, physical danger, and sexual abuse providing a primary focus. While not seeking to deny that such concerns may threaten the well-being of students, it is worth noting that these "perceived dangers" are presented as "broadly accepted," the "baseline risks" that must be addressed when discussing school e-safety. This creates a litany of preexisting online "problems" as defined by adults, which serves to restrict subsequent discussion concerning children's use of the internet and mute the voices of children themselves in subsequent debate. Thus, such risks are unproblematically listed in much child–e-safety literature. This also potentially fosters "diagnostic inflation" wherein new online risks, as defined by adults, are simply added to an ever-growing list of hazards that are presented as "self evidently problematic" and "potentially widespread." As Buckingham (2007, pp. 12–13) suggests, consulting the research literature on negative media impacts produces an "endless litany of evils." Yet this ignores the "state-of-the-actual" e-safety issues in schools.

This approach means that the "risk categories" are often ill defined. In the US Internet Safety Technical Task Force's *Enhancing Child Safety and Online Technologies* (2008), the simplistic discussion of risk results in the failure to consider how, for instance, pornographic material or cyber-bullying activities become labeled as such. The social construction of such concepts is treated in a largely unproblematic manner. Such limited consideration of the socially contested nature of risk not only disprivileges the views of children but also leads to contradictions in policy. For example, Becta's (2006a) policy guidance document *Safeguarding Children in a Digital World*, reinforces many of these adult risk discourses, while situating the discussion in a wider childcare framework, referencing policies such as The Children Act 2004, *Every Child Matters*, and *Safeguarding Children in Education*. Yet, contradictions abound as with the discussion of e-safety in relation to *Every Child Matters*, wherein there is no recognition that the goal of "staying safe" online, might conflict with two of the other key aims, namely "being healthy" and "enjoying life." After all, online risk taking might not only be enjoyable, but also part of a healthy development process. Ultimately much of the e-safety literature lacks cultural complexity, instead drawing upon a well-established list of adult-defined risks that often involve generalization and stereotyping.

The Depiction of Children in E-Safety Material

The manner in which children are represented in e-safety literature is often overly simplistic and polarized. Thus, there is a tendency to describe children as either innocent victims or dangerous perpetrators, as naive technology users or rational "digital natives." For example, Becta's *Signposts to Safety* (2007a, p. 9) notes that when using the internet "children are vulnerable and may expose themselves to danger." Yet, the e-safety literature also suggests that students can be a source of danger, as in *E-Safety* (Becta, 2005), where the discussion includes student activities labeled as "minor incidents" (plagiarism, illegal downloading, using another's password, and nuisance behavior with mobile phone technology), "incidents involving inappropriate material or activities" (such as downloading soft-core pornography, hate material, and drug or bomb making recipes), and "incidents involving illegal materials/activities" (in particular child pornography or extreme harassment online). Indeed it is suggested that for more serious misuse of the internet, police involvement could be necessary, effectively seeking to criminalize student behavior. Yet such representations lack social complexity that would allow a consideration that a student can be a victim and a perpetrator at differing times and, upon occasion, simultaneously. Rather there exists a polarization in the e-safety material of children as "innocent" or "dangerous." Furthermore, such conceptualizations rarely consider how such labels are socially constructed, culturally relative, and politically contested.

Similarly there is a contradiction in discussion concerning whether students are naive or savvy online. The following advice from the US-based Common Sense Media (2010, n.p.) website is typical of the first of these attitudes:

> As parents, it's up to us to help our kids understand the consequences of their actions and prepare them for the fact that the user name "FatGreenWizard"—which might have been cute in 5th grade—won't be so adorable at that first job interview.

Such narratives reinforce the role of parents, and often teachers, as protectors of naive, ill-informed children. In this context, there is little consideration that children may be knowledgeable technology users, who will learn to mold and redefine their online identities as required by future social situations. Yet, there is also a contradictory tendency to describe children as rational, savvy, and compliant in adult terms. Consider Becta's (2008a) expansive document *Safeguarding Children in a Digital World*, which contains a wealth of fictional case studies and e-safety dilemma cards, some of which are discussed in a "what happened next" section. The cards are intended for use by local "Safeguarding Children Boards," rather than students, and while they provide a useful focus in attempting to categorize risks and consider "who should be involved," there is an underlying assumption in the further discussion that the child at risk will behave in a "safe" risk-averse manner as defined by the adult supervisor. Thus the "dilemmas" are resolved as the children ignore spam, close inappropriate windows, avoid divulging personal details, suggest that their peers seek support, limit contact with online "friends," report abuse, and confide in parents or tutors. This is a somewhat naive position to adopt, suggesting that children will inevitably behave in a safe manner, as defined by adults. While it is recognized that "adult perception of the risk may be an overreaction" (Becta, 2008a, annex. D01), there is no questioning of whether all the dilemmas necessarily constitute a risk.

The tendency to treat children as a homogenous group in much e-safety literature fails to construct a complex, dynamic, socially grounded consideration of student's activities online. This failure tends to result in generalizations and polarized stereotypes, which do little to inform effective policy or practice. A temptation might exist to label younger children as innocent and naive, while applying the labels of dangerous and savvy to those who are older. Yet as has been suggested elsewhere (Hope, 2011), this would be a clumsy generalization, misrepresenting the social complexities. Rather tellingly, although children are variously constructed as victims or dangerous, naive or rational in e-safety literature, they are rarely labeled as citizens with rights. Thus, students tend to be treated as objects in e-safety literature, while little is mentioned of their rights in terms of privacy or freedom. Partly this is indicative of the lack of children's views in much of the e-safety material.

Consulting Children about Online Risk

Despite the need to listen to children, it would appear that young people's voices are marginalized in much school e-safety documentation. This is not an oversight, but reflects something of the power relations between children and adults. While in a few e-safety documents there are direct quotes from children, this does not necessarily equate to listening to what they have to say in a broader context. Indeed sometimes it is difficult to shake the feeling that such statements are included because they reaffirm the adult risk discourse. Perhaps it is unsurprising if policy or commercial documents do not seek to engage in a high degree of critical self-reflection, yet it does serve to construct a somewhat simplified view of student e-safety discourses.

Upon occasion the power relationships between adults and children are made explicit. For example, in the UK Council for Child Internet Safety's (2009, p. 14) *Click Clever, Click Safe* booklet, it is asserted that "we need to keep listening to parents and those who work with children" (but seemingly not the children themselves), while "we will continue researching how children learn and how they use the technology." Listening to children is disprivileged, as they become the objects of study, rather than the subjects of communication. While this treatment of children is undemocratic, it also does little to facilitate policymakers understanding of why children engage in online behavior that adults define as irrational or dangerous. With regard to school e-safety there is clearly a need for an inductive, child-directed, qualitative approach to risk research.

E-Safety Discourse and Social Control

Significantly, while children's voices are absent or muted in much e-safety literature, it is not uncommon to find the assertion that students need to be involved in the establishment of school e-safety policy. Yet, it is difficult to discern whether the intended outcome of such participation would be a widening of democratic practice, or merely the strengthening of existing social controls. Thus, the suggestion that students should be encouraged to contribute to the creation of e-safety policies through involvement in developing classroom rules and representation on the school internet policy team is followed in the Becta (2005, p. 27) booklet *E-Safety* by the following observation:

> Through this approach [participation], pupils will develop a greater understanding of the issues involved, and will feel more ownership and accountability for the policies. The ultimate aim is for pupils to take responsibility for their own actions when using the internet.

Insofar as the responsibilities of children are emphasized, it becomes difficult to avoid the conclusion that limited student participation is seen, at least partly, as a discursive, social control mechanism. Furthermore, it is subsequently suggested that student responsibilities might include "reporting any incidents of ICT misuse within school to a member of the teaching staff" (Becta, 2005, p. 27). It is not easy to reconcile open democratic communication with students policing their peers at the behest of the school. The subtext would appear to be that student participation would enable them to invest in adult imposed definitions of risk, socializing them into the dominant discourse, and encouraging them to regulate the behavior of their classmates. Indeed "a greater understanding of the issues involved" could be interpreted to mean children comprehending and accepting adult risk perspectives. It is never suggested in this document that adults need to "develop a greater understanding" of children's online-risk perspectives. Similarly, in Becta's (2009a) *Acceptable Use Policies in Context* report it is suggested that rather than restricting access to technology there is a need to encourage learners to develop safe and responsible online behaviors, to be able to protect (or possibly police) themselves. Consequently, it is asserted that "if children feel that their views and opinions have been considered, and can understand some of the issues affecting the decisions documented in Acceptable Use Policies, they may be more inclined to abide by them" (Becta 2009a, p. 21).

In the broader e-safety literature there are also commonly made calls for adults to engage in surveillance of children's online activities. For example, in On Guard Online's (2009, p. 17) e-safety booklet *Net Cetera: Chatting with Kids About Being Online*, which offers advice to adults, not insights from children, it is suggested that "if you're concerned that your child is engaging in risky online behaviour, you can search the blog sites they visit to see what information they're posting. Try searching by their name, nickname, school, hobbies, grade, or area where you live." Such recommendations are made without reference to individual rights.

Indeed a discussion of children's rights and issues such as freedom or privacy are conspicuous by their absence from e-safety literature. It might be expected that in guidance directly relating to the school curriculum and teaching e-safety such issues would feature. Yet in documents such as *Signposts to Safety* (Becta, 2007a, 2007b) a plethora of risks are discussed, alongside suggestions of where these issues can be addressed in school curricula in the subjects of ICT, Personal Social Health Education (PSHE), and Citizenship. Yet the suggestions seem primarily intent on reinforcing adult risk narratives and coercive practices, rather than asking more fundamental questions about the rights of children. It is difficult to escape Facer's (2012, p. 405) suggestion that "if children were the 'digital natives' of the internet, the adults had effectively exercised their power to colonise it."

The Commodification of Risk

Of course, it is not just the state that has served to craft the risk discourse relating to schoolchildren online. While school managers, governors, teachers, support staff, and parents all influence the social construction of the e-safety discourse at the microlevel, Buckingham (2007) also draws attention to what he labels as the "educational-technological complex," which includes quasigovernmental organizations, journalists, educational technologists, advisory services, teacher groups, researchers, commercial organizations, and non-IT related commercial interests. To explore the role of all of these actors with regard to school internet-risk discourses is beyond the scope of this chapter, yet the influence of commodification in the construction of national e-safety discourses is worth further consideration.

As individuals and institutions are held accountable for their own wellbeing and the safety of others, risk becomes commodified. School security devices are perceived as providing safeguards against risks, often regardless of actual impact. Following a liberalization of the market, high-tech and security companies have increasingly sought to exploit new opportunities in the UK and US compulsory-education sector (Kupchik & Monahan, 2006). Hence in recent years, schools have become burgeoning markets for private companies intent on selling the latest protection software, surveillance tools, and social sorting protocols. Thus commercial organizations have played a central role in introducing, as well as updating, school computer data bases (for monitoring attendance and facilitating threat analysis), internet tracking devices, plagiarism software, CCTV cameras (inside institutional grounds and on school buses), metal detectors, and biometric technologies (iris, fingerprint or face recognition devices). Such devices are intended to safeguard against an array of risks, ranging from the potentially life threatening to the mundane. Yet, as Casella (2010, p. 74) argues school security technology is not just concerned with well-being or control:

> While we may hear that security equipment is used in order to create a "safe environment" or, on the other hand, for "social control," these claims miss a significant aspect of school fortification: that the installation of security equipment in schools is foremost a corporate transaction led... by business people.

Risk monitoring and alleviation in schools is a multimillion-dollar industry. To not buy the latest risk amelioration technology may be seen as a failure of the school, ushering in the specter of blame. In short, it could be concluded simply that risk sells. As "new" threats become "recognized," private companies provide new technological solutions. Furthermore, security companies want individuals who already feel safe to continue to buy equipment. Risk can be mobilized as a commodity to suggest that it is never possible to do enough, to purchase sufficient security products or services, to truly safeguard what is held to be important. Thus, not only does risk alleviation technology promise safeguards, if only from blame, but also it acts as capital, signifying high expectations as well as a yearning for power and prestige. Security devices become life accessories, stylish accoutrements, and the latest "must have" tools. Such seductive narratives are often couched in terms of outmoded, negative approaches, future gains, and radical social-technological change. Consider the following advertising statement from the US software company iKeepSafe (2011):

> iKeepSafe's signature suite of services, Generation Safe™, allows schools to exceed the new [US e-safety] requirements, offering a paradigm shift through a digital literacy program that comes from a positive approach, rather than one based on fear mongering.

Significantly this software product is represented as offering a "paradigm shift" from fear mongering to a positive approach, which potentially safeguards not only students but rather an entire generation. Such claims reflect that "using security equipment and having it used on you is a sign of being forward-thinking and modern... technology (including security technology) is a sign of advancement" (Casella, 2010, pp. 79–80). It might seem strange given the consumer power of teenagers that companies do not pay more attention to children's risk narratives. The risk industry is built on adult risk perceptions, however. Possibly this reflects the reluctance of younger people to use their limited resources to safeguard against possible, future risks. Whatever the motivation, children's internet risk perceptions are disprivileged as commercial organizations attempt to enforce their views in a manner that ultimately promotes their products.

Risk as Property

If it is accepted that representations of children in e-safety documents lack social-cultural complexity and that children's views are disprivileged, as e-safety is mobilized as a tool of social control as well as a profitable commodity, then the question of what should be done arises. While a range of possible responses exist, key to any critical discussion is the privileging of children's perspectives and reconsidering the role of students in school risk discourse. One way in which to facilitate such processes is through reconceptualizing risk as property.

In a seminal article entitled "Conflicts as Property," Norwegian criminologist Nils Christie (1977, p. 1) argued that "conflicts have either been taken away from the parties directly involved and thereby have disappeared or become other people's property." He contends that the state acts on behalf of the victim, with the consequence that the victim might be completely pushed out of all proceedings, reduced to a mere trigger to the process (Christie, 1977, p. 3). Consequently, this person is a "double loser," not only a victim but also denied the rights to participation. Christie maintains that the victim loses his or her conflict to the state, starkly declaring that "he has a need for understanding, but is instead a non-person in a Kafka play" (1977, p. 8). If, within Christie's argument, the concept of risk replaces that of conflict then it draws attention to the situation that those at risk are not only potential victims but also have an important role as participants in subsequent processes, not necessarily of retribution or punishment but of (self-)healing and reconciliation. Yet, children have lost risk to the state, and their associated organizations, with the consequence that they often become shadowy stereotypes, powerless victims, who must be protected while their voices are ignored.

Thus, the online risks facing children should not be seen merely as something to be ameliorated by adults. Rather they are something that belong to children, about which they should have a say. While the Byron Review (2008) does not define risk in purely negative terms, recognizing that it may be the flipside to opportunity and that engaging in risky behavior may be part of the self-actualization process, this is not the same thing as declaring that children own their risk and have a right to be centrally involved in any subsequent process related to it. Redefining risk as property suggests that a reconsideration of how policy, research, and practice engage with children's risk perspectives is necessary. There is a need to move beyond treating children merely as a policy trigger and denying them the rights of participation.

Christie (1977, p. 8) believed that the loss of conflict to the state was also "a loss of pedagogical possibilities." Thus, redefining risk as property not only raises questions concerning ownership and children's involvement in related social processes, but also promises opportunities for learning. Indeed Meighan and Siraj-Blatchford (2003) assert that students learn about democracy through taking part in decision-making processes, thus failure to engage students in open dialogue concerning risk is a missed educational opportunity. These considerations gain in significance if we consider that risk is subjective and contested, as well as a normal part of the maturing process.

Conclusions

This chapter has made the case that the national school–e-safety agenda in the United Kingdom—as is likely elsewhere in the world—marks an attempt to impose adultcentric risk discourses, not just for the purpose of care, but also for coercive control and to facilitate commodification. As part of this process, children's risk perceptions that differ from adult viewpoints are frequently dismissed as ill informed, irrelevant, or irrational. Consequently, children often merely act as a policy trigger and are denied participation. Partly this reflects their legal status as minors, lacking full citizenship rights.

Reflecting upon this issue it can be argued that there is an urgent need for a critical approach in education technology writing that not merely focuses upon how digital artifacts are used, a so-called state-of-the-actual, but also that fully engages with broader, more complex social-political processes. After all, as Selwyn (2011a, p. 31) notes, "Many of the key questions surrounding education and technology are not concerned with issues of technology at all. Instead, they are related to wider questions of what education is, and what we want education to be." To this might be added the assertion that school national e-safety policy reveals much about what we want childhood to be and the role that we would prefer children to adopt. Viewing risk as property is one way in which this critical (re)privileging of the social context of education technology could be facilitated. Such an approach would underline the need for child-driven research that seeks to hermeneutically understand their worldview rather than dismissing it as ill informed. Ultimately, child e-safety and concern about school internet risks should not primarily be seen as matters of care, control, or even commodification, but more essentially a question of children's rights and a reflection of their subjugated position in society.

PART III

The Global Politics of Education and Technology

CHAPTER 6

"Empowering the World's Poorest Children"? A Critical Examination of One Laptop per Child*

Neil Selwyn

OLPC's mission is to empower the world's poorest children through education

—OLPC Mission statement

Introduction

The worldwide implementation of educational technology is underpinned by a wide range of interests and influences located at international, national, and local levels. Rather than being a globalized and uniformly determining force, any form of "educational technology" is itself dependent upon a number of social, cultural, political, and economic interests. Of course, this is not to deny that educational technologies are associated with some significant changes in education around the world. Yet anyone wishing to understand fully the nature and outcomes of educational technologies' use has to look far beyond the technical specifications and features of specific devices and gadgets. As has been argued by a succession of critical commentators over the past 30 years, educational technology needs to be described and discussed as a set of sociotechnical arrangements (Young, 1984; Kerr, 2003; Selwyn, 2012).

Focusing on the sociotechnical characteristics of educational technology inevitably raises questions of how, why, and in whose interests these devices and artifacts are used. In this manner, the use of digital technology in education—as with any aspect of society—is a profoundly political concern. There is a pressing need for critical scholars to attempt to look beyond the harmonious portrayals of educational technology that can often be found in popular, political, and academic discussions, and instead examine the areas of tension, contradiction, and conflict that underlie any instance of digital technology use in education. As was reasoned in the opening chapter of this book, educational technology is therefore perhaps best understood as an intense site of struggle across a number of fronts—from the allocation of resources and production of knowledge, to the maximizing of profit and political gain. As such, most of the questions that surround education and technology are the fundamental questions of education and society—that is, questions of what education is, and questions of what education should be. As the scope of these deceptively simple questions suggest, digital technologies are drawn inexorably into the global, national, and local politics of education—for better and for worse.

This chapter offers a critical reading of the politics of educational technology through a detailed examination of what many people consider to be the most significant global educational technology program of recent times. The "One Laptop per Child" initiative (OLPC) is one of the most ambitious, most publicized, and most lauded educational technology initiatives of the past 30 years. This is a program that claims to address many of the issues associated with the field of international development—not least child poverty, health, and universal access to education—yet at its heart has a universal agenda of promoting "technology-enhanced learning" across low-income and high-income contexts. Indeed, throughout the 2000s and into the 2010s, the goal of building and supplying a low-cost laptop computer for children and young people around the world has become a touchstone for progressively minded technologists and educationalists alike. Many people's faith in OLPC as a transformatory example of educational technology persists to this day. The initiative therefore offers an excellent case study through which to refine a critical sociotechnical analysis of educational technology. This chapter now goes on to examine the case of the OLPC initiative in detail—making sense of the rhetoric and the reality of one of the defining global educational technology programs of recent times.

The Technological Allure of OLPC

Many people find it difficult when discussing the OLPC initiative to look beyond the "laptop" device itself. Indeed, over the lifetime of the program the technological artifacts at the center of the "One Laptop" initiative have inspired many different descriptions. In monetary terms, the initially proposed "$200 laptop" soon became touted as the "$100 laptop" as the expected cost of its production began to fall. During the early years of the initiative, commentators referred playfully to the "Little Green Machine" and the "Children's Computer" as word spread of the devices' striking appearance and simplicity. Yet while prompting a number of different labels, OLPC has remained based around a disarmingly straightforward concept—that is, producing a low-cost, low-specification but highly durable personal computing device that can be handed over to children and young people around the world. As with many educational technology ventures, OLPC is often considered by its advocates to be an intuitive and commonsensical idea that transcends any future debate—what many people would describe as a technological "no-brainer." As Laurie Rowell (2007, n.p.) enthused a couple of years after the public launch of the program,

> Here's an outrageous idea, what if every child in the world could have a free personal laptop? Put some e-books on it, make it web-capable, and add a palette of media tools so children could work on creative projects. Wouldn't that be incredible?

While reassuring in its tone, the homespun enthusiasm that surrounds the project often serves to overshadow the precise nature and form of the OLPC initiative. From its inception, the program has been built around a central belief in developing and distributing devices that are designed specifically to bring networked computing (and, it follows, networked learning) to populations of children and young people who are otherwise living in disadvantaged conditions. While spokespeople for OLPC have constantly reiterated the claim that theirs is *not* a technology project, the outstanding feature of the initiative has been the innovative and ever-changing technical specifications of its computerized devices. At present the $100 laptop is in its third incarnation—the so-called XO-3 device that was developed after the program received a grant of over US$5 million from the multinational IT manufacturer Marvell to develop a low-cost tablet device built around low-power silicon chips. The XO-3 followed on from the "XO-2"—a flip-back, touch-screen "handbook" device, which can open flat to provide a square display supportive of writing, typing, and touch-sensitive input. Both these designs, in turn, followed on from the most iconic (and still predominant) OLPC device—the original "XO" laptop. While subsequent designs may have differed in their appearance, the XO laptop continues to embody the design principles and philosophies of the OLPC program. As such the XO remains the flagship technology of the program—especially in terms of the numbers of devices being used. Thus, before going on to consider the OLPC program in sociotechnical terms, it is important to first be clear about exactly what the XO is as a technological device.

The XO certainly stood out in terms of its appearance when introduced into the 2000s' consumer electronics marketplace. Described at the time of its launch as "a striking little green machine" (Naughton, 2005, p. 6), the XO gives the impression of being a sleek but durable child's plastic toy. Housed in rounded toughened plastic casing with a molded handle that resembled a lunch box, the most immediate qualities of the original XO were its size and color. This was a small lime-green and white device—weighing around 1.5 kilograms, and measuring little more than 22 centimeters square and 3 centimeters thick. When opened, a rubber-sealed keyboard, touch pad, and stylus were accompanied by a small pivoting display monitor. The idiosyncratic appearance of the computer was heightened by the inclusion of two extendible antenna "ears," designed to provide network connectivity to the internet as well as to other XO users within a radius of one kilometer.

Much of the initial excitement over the XO came from technology programmers and "hackers" who were drawn to this innovative technical design. Thanks largely to work from in-house developers and the Chinese "Quanta" computer company, the XO housed an impressive array of hardware features for a machine of its size and price—such as a microphone, camera, loudspeakers, "game controller" buttons, and USB and audio ports. Many of these technical features were intended to allow the XO to operate in inhospitable outdoor conditions. The laptop's keyboard, for instance, was rubber sealed and designed to be resistant to dirt and moisture. The plastic casing included built-in shock absorbers that were claimed to have been drop-tested successfully from heights of up to 15 feet. The display monitor was designed to offer low-power but high-resolution displays that altered appearance according to lighting conditions. The XO's screen could appear to be full color, pale color, or monochrome—thereby ensuring a readable display in even the brightest of conditions. Perhaps the most eye-catching components were the options for powering the XO—including windup hand-crank mechanisms and "yo-yo" pull-string power generators. These features, coupled with their nontoxic and fully recyclable design, were reckoned to make the XO computers "the most eco-green laptops that have ever been made" (Tabb, 2008, pp. 338–339).

Another technically appealing characteristic was the XO's innovative software design—in particular its reliance on open-source software and open-architecture hardware principles. Early incarnations of the laptop ran exclusively on slimmed-down versions of the Linux operating system coupled with a newly designed software interface titled "Sugar." This interface was intended to move beyond the usual "desktop" operating system design and provide users with an abstracted spatial navigation environment that supported navigation and collaboration via four levels of viewpoint labeled "home," "friends," "neighborhood," and "activity." Later versions of the XO offered a "dual boot" system that allowed the Microsoft Windows operating system and familiar "Office" software to also be used.

In the years following its launch, the XO has been roundly praised for its appearance, aesthetics, and overall quality of design. Indeed, most aspects of the OLPC program have reflected a high level of attention to design and detail that is often not found in mainstream computer production. While the XO hardware was developed by teams of in-house designers and small independent companies, the innovative design of the Sugar interface software was outsourced to the international product-design company Pentagram. OLPC therefore joined Pentagram's illustrious client list of Timex, Nike, United Airlines, and Swatch as part of the company's commitment to carrying out pro bono work for nonprofit organizations. Even otherwise-skeptical commentators were forced to concede the design qualities of the OLPC machines. As Linda Smith Tabb reported at the time of the laptop's deployment in US urban contexts, "The machines are truly revolutionary in design and almost every possible feature has been thoughtfully planned" (Tabb, 2008, pp. 338–339).

Aside from its high standards, the technical design of the XO is a particularly important aspect of understanding OLPC in sociotechnical terms—especially with regard to the values and agendas that have shaped the project from its start. Indeed, in terms of technical design, everything that has been described so far was influenced strongly by ideological values and intent. One recurring aim of the design of the XO was to produce an engaging and playful device that would appeal especially to young users. For example, the inclusion of the mesh-network antennae ears was intended to give the laptop an animalistic appearance akin to a rabbit (as well as giving it much-needed internet connectivity). Similarly, the software interface was designed deliberately to embody a philosophy of child-centered learning—placing the individual user at the center of a familiar environment that also promoted communal activity and collaboration. In all these ways, the XO devices were the result of a great deal of thought and attention. This was certainly not a profit-making "off-the-shelf" means of increasing levels of educational technology use around the world.

Unpacking the Sociotechnical Background of OLPC

On the face of it, then, OLPC could be seen as being an educational technology project almost beyond criticism—involving an innovative and thoughtfully designed piece of technology with the laudable aim of allowing children and young people to learn regardless of social circumstance. Indeed, many people within the educational technology community have been generally supportive of the promise to "create educational opportunities for the world's poorest children" through the production of a "rugged, low-cost, low-power, connected laptop with content and software designed for collaborative, joyful, self-empowered learning" (OLPC, 2010, n.p.). It is at this point, then, that we need to take a step back from the obvious allure of OLPC as a technological concept. What should be said about OLPC beyond its good intentions and innovative design?

First, it is necessary to place the program within a historical context. While undeniably ambitious, the OLPC was not the first initiative to seek to support low-cost computing for the masses (see Pal, Patra, Nedevschi, Plauche & Pawar, 2009). Even within the commercial confines of the consumer electronics market, US computer manufacturers were developing "low-cost, low-spec" computers throughout the 1980s—notably IBM's 1984 "PC Jr" model and subsequent competitor models such as the Tandy 1000. In terms of the development of low-cost computers for low-income countries, the Indian "Simputer" project was another prominent forerunner of OLPC. This attempt to develop a "Simple Inexpensive Multilingual Computer" also resulted in the nonprofit production of low-cost, open-source handheld computers with touch-sensitive screens. The Simputer was also accompanied by similar claims to the OLPC—as one commentator stated soon after its release, "This nondescript little computer may hold the key to bringing information technology to Third World countries" (Harvey, 2002, n.p.).

Over the past 15 years, a number of low-cost technological devices have also been produced for sale in developing regions. These included the production of "ultrabasic and ultracheap" computing devices based on Linux such as the Taiwanese "ASUSTek" computer and the Chinese "Lemote" laptop. Similarly, in terms of desktop computing, OLPC follows on from programs such as Brazil's Linux-based "Computador Popular" (people's computer), the Chinese "Rural PC," the "SuperGenius" Bharat PC, and the Apna PC—all relatively cheap devices aimed at extending access to computer technology to poor communities. These efforts reflect a long-held enthusiasm within the IT industry and professional technology community to establish "one-to-one" computing around the world—marked by the founding of an international group of high-profile technologists titled

"G1:1" (in full, "Globally, One Computer for One Person"). All of these precedents therefore raise a key point of interest—why has the OLPC initiative progressed so much further than these other ventures, and with what ultimate effect? Here, then, attention needs to be moved away from the technical aspects of OLPC devices and toward the nature of OLPC as a social and political project. As we will see, the OLPC program is as much a global political initiative as it is an educational technology initiative.

The Origins of OLPC

The team of academics and technology entrepreneurs behind the OLPC initiative came to the area of educational technology with considerable experience of similar ventures. The driving force behind the initiative from its start has been Nicholas Negroponte—a high-profile technologist and academic who was one of the founding members of Massachusetts Institute of Technology's (MIT's) prestigious "MediaLab" department. Along with MIT colleagues (and subsequent OLPC figureheads) such as Seymour Papert, Negroponte had been involved in an early computing project sponsored by the French government in 1982 ("Le Centre Mondial pour l'Informatique et Ressource Humaine"), which provided Apple II computers to Senegalese schools. Although relatively unsuccessful, the idea that children in developing regions of the world could benefit from the provision of computing resources was replicated in further projects during the 1990s and 2000s—in particular the involvement of MIT and Negroponte in the provision of internet-connected laptops to small groups of children in rural Cambodia, and the larger-scale distribution of laptops to seventh-grade students throughout the US state of Maine.

These early practical projects—and much of the intellectual work that occurred at MediaLab-sponsored conferences such as the "2B1" conference in 1997—were considered to provide an adequate "proof of concept" for this notion of one-to-one educational computing, prompting Negroponte's establishment of the nonprofit organization "One Laptop per Child Association Inc." After the official announcement of the organization and its intentions in January 2005 at the World Economic Forum in Davos, Negroponte presented a working prototype of the "Children's Machine 1" at the subsequent "World Summit on the Information Society" in Tunis. The choice of this high-profile audience for the launch of a still-to-be-finalized device resulted in considerable support being given to OLPC from across the international community, not least from the UN secretary Kofi Annan. Negroponte was celebrated in the *New York Times* as "the Johnny Appleseed of the digital era" (Markoff, 2005) and—despite appearing to break the prototype during the official launch—Kofi Annan himself welcomed the program as opening up a "new front" in the education of "the world's children."

In terms of actual production of the machines, the small-scale OLPC team worked hard during 2005 to gain formal support from other technology and media organizations—securing backing from the likes of Google, Nortel, News Corporation as well as hardware manufacturers such as AMD (Advanced Micro Devices), RedHat, and Quanta. After a further announcement at the 2006 World Economic Forum, the UN Development Program offered its formal endorsement and promised to act as a distribution agent for countries unable to purchase the minimum requirement. The actual production of the newly titled "XO" laptop then began in November 2007.

At this stage, the "business plan" for the OLPC program was a defiantly simple one. National governments that wished to participate were expected to each commit to the minimum bulk order of one million laptops. In turn, these governments would distribute the machines through their national educational networks to children and young people. In all cases, the distribution and implementation of the machines was to be conducted according to OLPC's five core principles, that is, "the kids keep the laptops, focus on early education, no-one gets left out, connection to the internet, and free to grow and adapt." Although a top-down model of state-directed distribution belied these bottom-up, open, and individually centered sentiments, the OLPC leadership viewed it as a necessary means of generating the volumes of investment needed to develop the technology. As such, only China, Brazil, Egypt, Thailand, and South Africa were considered initially to be worthwhile participants in the initiative. At this point, the OLPC program had clearly marked itself as a large-scale and politically astute social technology project—aiming to forcibly disrupt unequal patterns of access to digital technology in some of the world's largest but most deprived countries.

Initial Progress and Change

In practice, the progression of the initiative throughout the latter half of the 2000s was not as straightforward as these initial intentions would suggest. Above all, the relatively rapid introduction of the XO into the global IT marketplace of the 2000s provoked considerable criticism and resistance from other IT organizations—especially the XO's commercial competitors. This included Bill Gates's much-reported initial dismissal of the device in 2005. As Gates reasoned in a speech to a Microsoft Government Leaders' Forum, "if you are going to go have people share the computer...get a decent computer where you can actually read the text and you're not sitting there cranking the thing while you're trying to type." The XO's promise of low-cost internet connected computing therefore attracted a diversity of opposition within and outside the technology community—from the figureheads of worldwide technology corporations to individual developers and designers.

Perhaps the most significant challenge to OLPC to this time was the decision of Intel—then the world's leading producer of microchips—to produce its own low-cost, low-specification netbook computer for educational markets in developing countries. Titled the "Classmate" PC, this laptop was designed to retail to schools and students in low-income countries for between US$199 and US$299. Over a short period of time, Negroponte was forced to move from a position of defense (initially labeling Intel's intentions as "predatory" and "hurting the mission") to a position of consolidation—announcing a formal partnership between OLPC and Intel in 2007. However, this partnership was nullified after six months, with Negroponte demanding that production of the Classmate be discontinued before the two organizations could work together any further.

After this disjuncture, the Classmate program continued with laptops being produced and sold to governments around the world in conjunction with the Taiwanese manufacturer Asus. Part of the attraction to government purchasers of the Classmate products was their inclusion of Microsoft Windows and Office software, in comparison to the OLPC initiative's preference for bespoke open-source systems. Indeed, Microsoft's introduction at this time of a US$3 "Student Innovation" package of software to be sold in developing countries marked another direct challenge to the OLPC program. Many commentators were then only partially surprised by Negroponte's subsequent decision in 2007 to offer a version of the XO laptop with the dual option of open-source software and Microsoft Windows and Office software. This decision prompted widespread dismay from many of the OLPC's supporters—including the resignation for a time of the organization's "President for Software and Content," Walter Bender. Nevertheless, it highlighted a clear willingness to compromise the OLPC philosophy and the ideals of those involved in the face of commercial market-based concerns.

Throughout this period, actual sales of the XO continued to fall short of the initiative's projected numbers—with many of the early deals that appeared to have been secured with national governments failing to be completed. Most notably, contracts with Libya for 1.2 million XO computers and for 1 million units each with Nigeria and Thailand, all fell through. In an effort to counter this trend, further changes to the OLPC business model were then introduced—in particular the introduction of the "Give 1, Get 1" (GIGI) scheme during the Christmas season of 2007. Here, North American consumers were allowed to purchase an XO computer for US$399, with this cost covering the donation of an additional laptop to specified programs in low-income countries. As a further inducement, customers were able to have the donated computer considered as a tax-deductible charitable contribution. This subsidizing of XO donations by the North American market saw nearly 84,000 donations being made. Tellingly, a second GIGI scheme the following Christmas through the online retailer Amazon saw only 12,500 laptops being sold. Nevertheless, this renewed momentum saw XO laptops distributed to countries such as Haiti, Afghanistan, Mongolia, Ethiopia, and Vanuatu. This geographical spread of the XO continued with the sale of thousands of laptops to the New York City Department of Education, as well as the education departments of Chester County in Pennsylvania, and Birmingham Alabama. At this point, the OLPC initiative was certainly increasing its global reach, although without necessarily achieving the levels of saturation that had been promised initially.

Recent Developments

The OLPC initiative has latterly found itself continuing to be one of the most substantial—and certainly most visible—global educational technology projects of recent times. The program has continued to attract considerable amounts of support and publicity into the 2010s. The donation of XO computers were part of aid efforts in the aftermath of the 2010 Haiti earthquake, as well as being deployed into other high-profile humanitarian zones such as Iraq, Gaza, and Afghanistan. In commercial terms, the concept of the XO is considered to have hastened the emergence of the low-cost "net book" market in Western countries. Conversely, politicians have continued to laud the initiative as an example of innovative international development—to the point of calls "for the OLPC program to be designated by the UN as a new Millennium Development Goal" (Tabb, 2008, p. 337).

In this respect, the OLPC could be judged to have been one of the most successful educational technology programs of recent times. The initiative has been implemented in a number of South American countries, with governments in sub-Saharan Africa also participating. This has seen the introduction of over one million machines into Peru and Uruguay, with smaller amounts in the countries such as Ghana, Argentina, Columbia, Mexico, and Rwanda. Coupled with the machines that have been introduced through loss-leading pilot programs and the GIGI donations, this means that OLPC computers can be found in over 20 countries from Nicaragua to Nepal. The initiative has certainly prompted changes in the patterns of educational technology use in some of these countries. In Uruguay, for example, XO laptops are at the heart of the "Conectividad Educativa de Informática Básica para el Apredizaje en Lánea" (CEIBAL) initiative—reckoned to be the first national program to achieve a one-to-one ratio of primary school students to computers.

That said, these levels of adoption have failed to match the initial expectations and proclamations of Negroponte and his team. As Yujuico and Gelb (2011, p. 50) concluded, "If the criterion for success was admiration for an innovative concept, the OLPC project would be an unqualified triumph...however, if the criterion was achieving its sales goals, the project would have to be judged a failure, despite some recent glimmers of progress." For many commentators within the technology community, one key failing of the initiative has been the stabilization of the actual price-per-laptop at a level approaching US$200. Coupled with the added "financial burden" of maintenance, technical support, and other aspects of program maintenance, the OLPC devices clearly cost far above the mooted price of US$100 (Streicher-Porte et al., 2009). Yet despite these issues of price and penetration, OLPC remains a beacon project for many educational and technological commentators—seen to offer clear proof that digital technology can be an integral element of a transformative agenda in the field of international development. As de Bastion and Rolf (2008, p. 31) conclude with regard to the continued rollout of the XO machines in sub-Saharan Africa,

As an integral part of a robust overall strategy, it is indeed correct to give children in Ethiopia a laptop...It may seem ironic to distribute emergency aid and computers at the same time, but it is one way of breaking the endless cycle of dependency. The true madness would be to underestimate the lasting value of the learning which ICT4D can additionally deliver.

Unpacking the Values of the OLPC Program

As this brief overview of its progress suggests, OLPC is certainly not a straightforward technology production and distribution program. Indeed, seen in terms of a theoretical focus on the "social shaping" of technology, the OLPC initiative is better understood as being driven at all stages of its development by a complex set of interests, values, and guiding agendas. As such, the idea of putting an XO laptop in the hands of every child in the world clearly has been—and continues to be—informed by a set of accompanying ideological interests and agendas. In this respect, OLPC is no different to all of the other examples of educational technology considered by other contributors to this book.

The notion of educational technology as an ideologically driven process is not lost on those involved in the OLPC initiative. As Nicholas Negroponte has himself reasoned, "We're not building an empire, we're building a movement" (Negroponte cited in Hamm & Smith, 2008). Thus, as Ananny and Winters (2007, p. 107) continue,

> We suggest that this and other projects be critiqued not only in terms of their technological feasibility, economic rationales or models of education but, more fundamentally, in terms of the ideologies they intend their users to enact. [Even] the OLPC's interface guidelines...serve— intentionally or otherwise—as powerful signals to policy makers, cultural critics and local communities of the particular ideologies intended to be enacted by the XO's users.

In this manner, we now need to move beyond our initial descriptions of the OLPC as a set of artifacts (e.g., the XO devices and their software designs) and as a set of practices (e.g., the design decisions of OLPC, its partnering organizations and community of open-source developers). Instead, we now need to consider the OLPC initiative as embodying a set of values, and approach the XO laptops as "sites in which designers, users, policy-makers and evangelists of all stripes perform ideology—explicitly or otherwise" (Ananny & Winters, 2007, p. 117). From this perspective, a number of different ideological assumptions can be identified as having underpinned the OLPC implementation to date.

First is the assumption that the XO laptops offer a means of achieving significant social, economic, cultural, and political change in developing regions and countries. Indeed, much of the popular appeal of the OLPC project stems from the grandiose "noble dream" (Rowell, 2007) that informs many of the initiative's actions and activities. Behind the impressive proclamations relating to "the idea that universal laptop computer use will revolutionize the world for the better" (Luyt, 2008, n.p.), lies an aggressive modernization agenda, similar to many of the educational projects in the field of "ICT4D" (Information Communication Technology for development). Indeed, much of the impetus behind the OLPC initiative stems from a belief that enhanced access to technology can lead to a range of educational, environmental, and societal-related improvements. As Nicolas Negroponte has asserted,

> The more people that are capable of rational, critical thinking, the better the world will be. The more they have access to knowledge about the rest of the world, the better the world will be. This is probably the

only hope—I don't want to place too much on OLPC—but if I really have to look at sort of, how to eliminate poverty and create peace and work on the environment, I think—I can't think of a better way to do it. (Negroponte, cited in Dotsub, 2007)

As this bold statement implies, many of these societal benefits are seen as achievable through the stimulation and support of technology-enhanced learning directly and indirectly. In this manner, OLPC is built deliberately around a context-free "liberation-from" model of social change. As Kullman and Lee (2012, p. 47) observe, "For the developers of OLPC, capability and liberation are achievable through increasing the ability of individual children to safely ignore and transcend their immediate circumstances." Such is this conviction that there has been increasingly frenzied rhetoric of securing the distribution of OLPC devices to children regardless of their circumstance, with Negroponte even making the proposal at the Social Innovation Summit in New York for a mass air-drop of tablet computers into remote villages from helicopters (Bajak, 2012)

Alongside this unshakable faith in the primacy of individual agentic action, it is also important to note that the XO laptop has been built around a very specific set of assumptions about education and learning. From its start, OLPC has been positioned deliberately around a set of social-constructivist learning principles common to most MIT MediaLab projects. Here the initial involvement of the prominent MIT professor Seymour Papert in OLPC is an important factor in understanding the values underpinning the program. Papert's well-known refinement of social-constructivist learning theory into the notion of "constructionism" during the 1970s and 1980s provides a clear underpinning principle for the technological and pedagogical design of the XO laptops. Through constructionism Papert proposed that learning best takes place when individuals are engaged in socially rich informal learning environments where they can create computational objects and systems that act as concrete representations of their cognitive development. As such, constructivist principles have been explicitly "built-in" to most aspects of the XO design—from the Sugar interface to the anthropomorphic network antennae. Indeed, the early label of the Children's Machine 1 for the XO laptop deliberately referred back to Papert's 1996 book on constructionism and computers entitled "The Children's Machine."

Allied to these beliefs in learner centeredness is a guiding value throughout OLPC of networked individualism and a belief in the self-determining power of the individual. Indeed, the constructionist ethos is built around an individualized notion of learning—with the individual learner responsible for coordinating and directing his or her own educational experiences. This philosophy is reflected, for example, in the positioning of the OLPC initiative around "a particular model of children as agents of change and networks as the mechanisms of change" (Ananny & Winters, 2007, p. 107). Politically, then, the OLPC initiative moves beyond supporting the increased engagement of individuals with learning, to wider issues of supporting individuals in taking complete control of the process of education. Thus, as Michael Klebl observes, the OLPC initiative does not seek to support change through the enhancement of education systems or education institutions—"instead of traditional methods for improving an educational system like building schools, spreading textbooks, reforming the curriculum or educating teachers, self-determination of the children themselves is at the midst of this educational reform, leveraged by a technical device" (Klebl, 2008, p. 280). As such, it is unsurprising that the initiative has been increasingly presented as an "educator-free" model of learning—as Negroponte was recently quoted as arguing,

> There are about 100 million kids without schools, without access to literate adults, and I would like to explore a way to get tablets to them in a manner that does not need "educators" to go to the village. (Negroponte, cited in Bajak, 2012, n.p)

Of course, a belief in individualized self-empowerment has long run throughout the field of educational technology, reflecting an implicit (and sometimes unconscious) "romantic individualism" among many technologists that positions individual technology users as "inherently expressive and self-transforming" (Luyt, 2008). Yet there is also a clear anti-institutional element to the OLPC philosophy that is less common to other educational technology projects and programs. Despite relying on national school systems to facilitate the distribution of XO laptops to children and young people, there is a distinct antischool sentiment to the OLPC project. As Laurie Rowell (2007, n.p) reported at the time of production of the first incarnation of the XO laptop,

> Walter Bender has been clear in saying that education could benefit from a paradigm that allows more critical evaluation from people at all levels, and he's frank in suggesting that the traditional school hierarchy is a barrier to quality improvement. In his words, the education community, because of the way school and (perhaps more significantly) school systems are structured typically top down, tends to suppress the spread of best practice as it is developed bottom-up in the classroom.

While concerned with wider social issues such as the deinstitutionalization of education, a further set of philosophies that have underpinned the OLPC initiative since its inception are more explicitly technologically driven—what Andrew Brown (2009, p. 1152) has labeled "the fetishizing of technology, and the pursuit of access as a social project in and of itself." These values were perhaps best expressed by one of the early slogans adopted by the community of programmers responsible for the initial development of the XO. Thus it was stated succinctly, "Not every child in the world has a laptop. This is a bug. We're fixing it" (cited in Klebl, 2008, p. 280). As this melding of programming logic and social welfare suggests, the OLPC initiative has gained considerable momentum from its positioning as a collective effort on behalf of the technology community to develop technically sophisticated and exciting machines. This is noticeably the case in the high-profile alignment of the XO devices with open-source principles. Of course, the notion that the XO hardware and software is "open" to user reconfiguration and improvement chimes with the constructionist and constructivist learning theories outlined previously. Yet the open-source label has also been valuable in giving the XO a technologically "cool" cache that some critics argue has gone some way to obscuring—or even overcoming—any criticism of the devices' clear technical limitations. As Brown (2009, p. 1168) concludes, "Though it is claimed that this is an education not a technology project, the development of the laptop, rather than the principles of its use, have been to the fore."

Of course, this "core" philosophy of following open-source principles belies the OLPC program's almost ruthless commercial and political pragmatism when it has come to achieving its aims. As described above, the history of the OLPC initiative has been characterized by an ability to broker deals and partnerships with previously conflicting interests and organizations. This can be seen in successive arrangements with commercially hostile organizations such as Intel and Microsoft, as well as the maintenance of partnerships with supranational and intergovernmental organizations such as the World Bank and the United Nations, alongside corporate partners such as Google, Amazon, Citigroup, and eBay. At the time of writing, plans to partner with the Walmart supermarket chain were being mooted by the OLPC management team. As such, a clear philosophy of political adaptability and pragmatism runs throughout many of the OLPC team's actions. As Nicolas Negroponte reasoned when defending the decision to offer a dual open-source/Microsoft product, "It's like Greenpeace cutting a deal with Exxon. You're sleeping with the enemy, but you do it" (cited in Hamm & Smith, 2008).

OLPC: Toward a Critical Perspective

The scale of these ambitions—and the aggressive and often self-important manner in which they have been pursued over the past ten years—has understandably begun to attract a burgeoning critical commentary. Yet it is telling that popular discussion of OLPC has, for the most part, taken place in an empirical vacuum. Despite some hagiographic "evaluation" and "assessment" studies, there has been little tangible evidence of sustained effectiveness and outcomes. The few independent studies that have been conducted of XO implementation have raised doubts of any substantial changes taking place in situ. As the authors of one evaluation of the OLPC implementation in South America concluded,

> Our interviews and observations in Paraguay suggest that XO use there is stratified, with a minority of youth making use of the XOs in creative and cognitively challenging ways, and a majority using them only for simpler forms of games and entertainment. We also found that the children who are already most privileged socially and economically tend to make use of the XOs most creatively. Thus, independent XO use by children might exacerbate divides rather than overcome them. (Warschauer & Ames, 2010, p. 44)

This lack of empirical "evidence" stems, in part, from the OLPC program's dismissive stance against the need for evaluations and pilot studies to be conducted. As Nicholas Negroponte reasoned in a 2009 speech entitled "Lessons Learned and Future Challenges,"

> I'd like you to imagine that I told you "I have a technology that is going to change the quality of life." And then I tell you, "Really the right thing to do is to set up a pilot project to test my technology. And then the second thing to do is, once the pilot has been running for some period of time, is to go and measure very carefully the benefits of that technology." And then I am to tell you what we are going to do is very scientifically evaluate this technology, with control groups—giving it to some, giving it to others. This all is very reasonable until I tell you the technology is electricity, and you say, "Wait, you don't have to do that." But you don't have to do that with laptops and learning either. The fact that somebody in the room would say the impact is unclear is to me amazing—unbelievably amazing.

It could be argued that Negroponte's bullishness reflects an imperviousness to criticism that leaders of large-scale global projects undoubtedly require to succeed. Indeed, a strong conviction and sense of righteousness pervades much of the commentary that surrounds OLPC. Yet as even its most ardent supporters acknowledge, the enormity of the project has left the OLPC program falling short of its much-publicized ambitions. For example, while the initial stated target of the Australian OLPC program was the provision of 400,000 laptops to children in remote regions, the actual delivery achieved since January 2008 has been closer to 5,000 machines. Through instances such as this, OLPC has begun to attract growing criticism in contrast to the initial wave of positive support and enthusiasm. Indeed, the XO laptop itself was reported by the *New York Times* to now be regarded by some sectors of the NGO community as "the emblem of the failure of technology to achieve change for the better" (Strom, 2010, n.p.). The suggestion can be made, therefore, that the OLPC initiative has been thwarted by a set of mitigating issues that face any large-scale educational technology program. These issues are therefore worth considering in more detail if we are to extend the example of OLPC to other examples of educational technology around the world.

First is the contention that the XO devices and the wider ambitions of the OLPC program are simply inappropriate for the contexts in which they are being implemented. In particular it is argued that the XO machines have been "designed in a lab-centric rather than need-oriented paradigm," therefore failing to fit with the realities of the developing countries and poor regions where they are being implemented (Pal et al., 2009, p. 61). It certainly could be argued that the OLPC initiative is founded upon exaggerated expectations of the vitality of laptop computing outside of the developed world. As John Naughton (2005, p. 6) has argued, distribution of XO devices to communities in sub-Saharan Africa raises significant questions of appropriateness, not least "whether the folks who wrote it have any understanding of what it's like to live in a society where the average income is less than $2 a day and the notion of children's rights is as theoretical as time travel."

Taking these concerns further, it has been reasoned that machines worth upward of US$200 are of considerable value in most of the countries where they are being delivered. From a practical perspective, this has prompted a reticence among some children, young people, and teachers to use the devices in fear of damaging or breaking them. As Warschauer and Ames (2010) found, some communities of XO users have also encountered difficulties in meeting the costs of running the machines and then ensuring the provision and maintenance of basic infrastructure. Another salient issue along these lines is that many children are simply unreliable "owners" of laptop computers. As Warschauaer, Cotton, and Ames (2011, p. 71) observed,

> When children own their own laptops and are responsible for maintaining them, over time many of them break down and go unrepaired. Moreover, the poorest children and families are most likely to be unable to repair their laptops. This results in a situation, confirmed by our classroom observations, in which large numbers of students do not have working laptops.

Indeed, in many different contexts the XO machines have proved difficult to repair and to find replacement parts. In practice, some of the key components of the XO-1 laptop (such as the rubber membrane keyboards) have been found to quickly wear out and render the machines useless. Studies of the OLPC projects in New York City and Alabama found that large numbers of laptops were broken or otherwise unusable within the first 20 months of implementation. Similarly, in Uruguay it was reported that more than half of the XO machines that were out of commission were determined to be unusable due to breakage. As Warschauer and Ames (2010, p. 41) continue, "Earlier, Papert claimed that 'an eight-year-old is capable of doing 90 per cent of tech support and a 12-year-old 100 per cent.' This may well be true in theory, but in practice large numbers of XOs go unrepaired."

A set of concerns is also emerging with regard to the "goodness-of-fit" between the OLPC program and the school contexts within which the laptops are being deployed. One case study of XO laptops in Indian schools, for example, reported challenges ranging from infrastructural issues relating to reliable power supplies and keeping laptops charged, to the lack of structured support for teachers. It also highlighted the shaping of the laptop use around the existing micropolitics of school time, school ownership of resources, preferences for teacher surveillance, and paired (rather than individual) working patterns (Padmanabhan & Wise, 2012). As another investigation of laptop use in Peru found, students already familiar with video games were largely disinterested in the "foreign experiences" of the XO laptops, while many teachers perceived the devices to conflict with their preexisting arrangements. As such, these researchers concluded, the OLPC devices did not appear to fit with students' or teachers' contexts—"this particular sociotechnical system evidently arrived without any strategy for embedding itself in the daily practices of a different set of agents" (Villanueva-Mansilla & Olivera, 2012, p. 185).

These practical technical limitations are compounded by a set of wider moral issues—not least the appropriateness of directing funding and resources toward what is essentially a global educational technology experiment. As Andrew Brown (2009, p. 1152) concludes, "With even small amounts of money able to make a distinct difference to life chances in desperately poor parts of the world, through, for instance the provision of fresh water and vital medication, this effort is misplaced." These criticisms are especially acute with regard to the "goodness-of-fit" between the OLPC program and the nature of education systems in developing nations. For instance, as Larry Cuban observed, many of the guiding philosophies behind OLPC could be considered to be "naïve and innocent about the reality of formal schooling" (cited in Markoff, 2006). The educational and pedagogical merits of the OLPC philosophy have therefore been challenged from a number of perspectives—not least the lack of testing and research into the educational assumptions that underpin the initiative. For instance, the philosophy of not encouraging the sharing of resources within communities has been criticized in terms of restricting any benefits of the program to a minority of children and young people (James, 2011). More broadly, as John Naughton (2005, p. 6) queried at the launch of the program,

> It is an article of faith that giving kids computers is a way of aiding their learning... [The OLPC initiative] is thus rather grandly contemptuous of mundane questions such as whether there is any evidence that giving kids computers is educationally better than giving them books, hiring more teachers or building more schools—or even paying families to send their kids to school. For Papert—and his MIT colleagues—technology seems to be the answer, no matter what the question.

Clear divisions can therefore be traced between the OLPC program and the educational systems that they seek to initially work within but ultimately intend to then work around. In particular, it could be argued that the OLPC initiative suffers from a conceptual tension in viewing individual children as the principal sites of change while also using the principal mechanism of change as the networked structures of national school systems (Ananny & Winters, 2007). Indeed, the deliberately provocative strategy of giving laptops to individual children and young people has prompted considerable unease among those with vested interests in the continuation of formal educational institutions. As the general secretary of the Peruvian "Unified Union of Education Workers" was reported to argue, "These laptops are not part of a comprehensive educational, pedagogical project, and their usefulness is debatable" (Luis Munoz Alvarado cited in Hamm & Smith, 2008).

Of course, the story of OLPC "is not simply about the failure of good intentions via inadequate design" (Philip, Irani & Dourish, 2012, p. 4)—this is as much a story of the politics of global education reform as it is about the development and implementation of the OLPC digital devices. First and foremost, the ambition of OLPC to import (and many would argue impose) a homogenous set of "other" principles and values into a diverse range of countries and contexts around the world has raised concerns over the program's cultural insensitivity and neocolonialist approach. As Andrea Chan (2012) argues, OLPC typifies the "universalist" underpinnings of contemporary digital technology that allows a vision for digital connection generated by cosmopolitan techno-elite, to speak for and represent the "global" rest. For some critics, then, the OLPC program is an extension of earlier colonialist interventions into the regions of South America and sub-Saharan Africa. As technologist Guido van Possum has argued, "I have thought for a while that sending laptops to developing countries is simply the twenty-first century equivalent of sending Bibles to the colonies" (cited in Brabazon, 2010). In particular, the "one-size-fits-all" model of OLPC in terms of technology, pedagogy, and business has been widely criticized. As Allen (2012, p. 207) concludes, OLPC "is entrenched in Western values and ideals and thus influences developing countries and the Indigenous communities within those developing communities in a culturally negative way." While extreme, these criticisms are reflected in practical aspects of XO use in local contexts—in particular the OLPC model of open-source development of diverse "local" content. As Linda Smith Tabb (2008, p. 347) observes,

> Since most of the translators for the project are volunteers, it seems improbable that all of the various languages will be able to be used for the XO laptops. This is a concern even in countries such as Haiti—where Kreyol Aiysyen co-exists with French—and the Andes of Peru—where Quechua co-exists with Spanish—where linguistic recolonisation is at risk if laptops do not enable use of either language. In these cases, the Green Machine does not so much threaten Americanization, but cultural absorption by polities of larger scale in closer proximity.

Echoing the concerns of Hall (in ch. 9 of this book), strong criticisms can also be made of the largely obscured issues of exploitation and domination that underpin the OLPC model. In particular, the aggressive posturing over the reduction of the price of the device to the level of US$100 and the popular promotion of the programmers, marketers, and users of the XO as "heroic actors" obscures a number of issues of exploitation that lie beneath the popular promotion of the heroic actors of the project. Most notable among this exploitation is the reliance of the project on "the standing reserves of feminized Asian labor that manufactures the XO laptop" (Philip et al., 2012, p. 11).

One final criticism is a misplaced confidence in the ability of a global technology project such as the OLPC to be free from political influence and interference while still dealing with national governments and multinational corporations. From the perspective of the OLPC leadership, the decision to focus the program on state educational systems was largely strategic and self-serving. Indeed, it has been acknowledged that focusing the OLPC program in terms of education and learning has been a convenient means of "translating" the XO laptop "into ways that fit with the mission" of the governments and state bureaucracies that the OLPC team needed to work with in order to achieve maximum coverage (see Luyt 2008). Indeed, Negroponte (2007) has been explicit in the role of notion of the "educational" $100 laptop as a "Trojan Horse" tactic to get the technology into the hands of children and young people.

Yet in taking this pathway, OLPC has shown a considerable lack of political realism in its dealings with national governments and multinational corporations. As has been suggested throughout this chapter, from its inception onward the OLPC initiative has been mired in the politics of international relations and of international commerce. Since its launch, the OLPC leadership has failed to find political ways of countering the continued reluctance of governments willing to commit to the required mass orders of machines. As one commentator observed four years after the high-profile launch of the initiative, "after years of deal-making and political machinations, it is still only making relatively slow progress" (Johnson, 2009, p. 5). This political intransigence was illustrated with the deployment of XO machines to Iraq in the aftermath of the second Gulf war. Although much heralded at the time as an instance of OLPC bringing technology and education to otherwise deprived contexts, in reality Iraqi use of the XO laptops was minimal. As Warschauer and Ames (2010, p. 36) note, "The U.S. government bought 8080 XOs for donation to Iraq, but they never reached children's hands; half were auctioned off to a businessman in Basra for $10.88 each and half are unaccounted for."

Of course, political intransigence and compromise is part and parcel of international relations. Yet as far as the OLPC team seems concerned, these barriers have been mostly unexpected. As Negroponte conceded in 2007, "I have to some degree underestimated the difference between shaking the hand of a head of state and having a check written, and, yes, it has been a disappointment." As Linda Smith Tabb (2008, p. 339) reasoned in response,

What is most striking about his statement, besides the obvious arrogance it takes to assume that a deal with a head of state could be so easily facilitated, is the disregard for the speed of a liberal democratic process, which is usually very slow. The main type of leader who might be able to make good, and fast, would be one not interested in a completely demo-cratic process of decision making and consensus building. So, in Latin America, the traditional home of the caudillo, decisions seem to be made at a much swifter pace than in the rest of the world. The campaign at this juncture could have been renamed "Una Computadora por Niño."

These are certainly harsh criticisms. Yet the ease with which the OLPC program brokered deals with world leaders such as former president Tabaré Vázquez of Uruguay, former president Gaddafi of Libya, and former presi-dent Olusegun Obasanjo of Nigeria certainly suggests a political expediency (and possible lack of concern for ethical and moral consistency) when pursu-ing the aim of getting the XO laptops into the hands of schoolchildren.

The political complexity of the OLPC program's dealings with national governments is illustrated by the ongoing failure of the initiative to be adopted in India. Despite placing a great deal of emphasis on the need to establish the program in the country (Negroponte was once quoted as saying, "India is the largest market for us, and I had to be there"), there have been numer-ous public denouncements from the Indian government to the advances of the OLPC team. Government officials argued in 2006, for example, that "India must not allow itself to be used for experimentation with children in this area" (Mukul, 2006). Sudeep Banerjee, head of the Indian Ministry of Human Resources Development, branded the idea "pedagogically suspect," and suggested that "classrooms and teachers were more urgently needed than fancy tools." As another official from the Human Resource Development Ministry concluded, "It would be impossible to justify an expenditure of this scale on a debatable scheme when public funds continue to be in inad-equate supply for well-established needs" (Mukul, 2006).

Such reactions are not attributable solely to a skepticism among Indian politicians about the social and educational merits of the XO laptops, but also reflect a general wariness of grand technological solutions from exter-nal Western organizations. Notably, OLPC undoubtedly suffered from Negroponte's prominent involvement in a previous project to establish a satellite "MIT Media Lab Asia" in India, which ceased despite significant amounts of initial funding from the Indian government. Also significant has been the Indian government's desire to convey its political ambitions to be seen as an emerging superpower capable of supporting its own tech-nology projects. As the Human Resource Development minister stated at the launch of a proposed Indian-built US$35 tablet computer, "The solu-tions for tomorrow will emerge from India" (Kapil Sibal, cited in BBC News 2010).[1] Against this local political context, the assumed global appeal of the OLPC program has understandably failed to take hold.

Conclusions

The case of OLPC encompasses a complex knot of issues and agendas—connecting design issues with matters of social justice on a global scale, alongside cultural assumptions about "development" and the role of technology in effecting social change (Kullman & Lee, 2012). All the criticisms of the program outlined during the latter sections of this chapter should not detract from the many positive outcomes that have certainly arisen from the OLPC initiative so far. These include the foregrounding of the issue of low-cost computing onto the global political stage, as well as the many considerable advances in the technical development of low-cost computing components that have derived from the development of the XO devices. Yet, as Warschauer and Ames (2010, p. 46) note, "There are important differences between a research-oriented development effort and a large-scale international campaign involving the production, distribution and use of millions of educational computers." It is here that the gulf between the grand ambitions of the educational technology community and the realpolitik of world economics and world politics are laid bare. Through its consideration of the OLPC initiative as more than a well-designed and well-intentioned technological device, this chapter has been able to further explore some of the key themes that have underpinned the increased digitization of education over the past three decades or so—not least issues of power, politics, and ideology.

From even this brief discussion of the program, the complexities of OLPC are obvious. What might appear at first glance to be an innovative and ambitious educational initiative has in fact been shaped by a number of mitigating factors. These include the professional backgrounds and beliefs of its founders; the educational, economic, and ecological values driving the design of technology hardware and software; and the complications of intervening in commercial markets and "selling" not-for-profit technologies to state purchasers. As such, this chapter offers a rich account of the politics of educational technology. What may appear to be an uncontroversial international development project has, in practice, proved to be a site for a number of ideological conflicts. These conflicts include the privileging of the assumed power of individual actors and market forces over the governance of national governments, as well as the deinstitutionalization of public services such as schools and schooling. Far from being a benign force for "good," the OLPC "mission" of putting low-cost, brightly colored digital devices "into the hands of children" has been driven by complex political struggles and conflicts. As such, OLPC highlights a number of salient issues that should be taken forward into general discussions of the politics of education and technology.

First and foremost, the OLPC program highlights the need to balance any focus on the design and development of technology with consideration of social, political, cultural, and economic contexts within which technology use takes place. It could be argued that many of the "unexpected" setbacks now being faced by OLPC implementation in various local contexts relate back to the overtly technicist nature of the project and the excessive faith put into the XO technology itself. Indeed, beyond the understandable criticism of the OLPC program that "the hackers took over" (Edith Ackermann, cited in Hamm & Smith, 2008), lies a willingness among many people within the educational technology community—not least Negroponte himself—to approach the social issues that are supposedly being addressed through technology programs in largely technical terms. In particular, many of the OLPC actions appear to have been informed by a prevailing view of social change as a form of programming-orientated problem—that is, as a logical series of "bugs" in a system that need to be fixed. As Michael Klebl (2008, p. 280) reasons, OLPC therefore could be said to "represent an interpretation of educational expansion solely as a technical issue to be solved like a programming mistake. An inexpensive, connected, and robust laptop personally owned by every child provides the ability to learn and progress, especially for children in developing countries."

Above all, then, OLPC stands as another reminder of the tensions between global technology solutions and local contexts of implementation. Despite the technical elegance and apparent political simplicity of the OLPC business plan, it would seem that no amount of charismatic leadership, strategic lobbying, and technological sophistication can impose globalized change and transformation onto whole societies or national education systems. Perhaps most importantly, OLPC reminds us that "there is no such thing as 'actor-free' dissemination or reception, lending or borrowing, export or import" (Tabb, 2008, p. 345). As with all the examples of educational technology discussed in this book, the grand global ambitions of OLPC are entwined with the mundane realities of the local educational settings and contexts in which they seek to be located (see also Cervantes, Warschauer, Nardi & Sambasivan, 2011). It is this thought—above all others—that needs to be taken forward as a basis of any attempt to advance the field of educational technology toward more equitable ends.

Notes

*This is a revised and updated version of "One Laptop per Child: A Critical Analysis"—Chapter 7 (pp. 127–146) in N. Selwyn (2013), *Education in a Digital World*, London, Routledge. The author would like to thank the original publishers for their permission to republish this work.

1. The Indian US$35-tablet scheme is itself an interesting case study of the complex politics of high-tech "big-gesture" programs. Eventually settling on a US$50-Android-powered tablet manufactured in China, the program has been marred by criticisms of business mismanagement and unrealized targets, with serious doubts being raised over the viability of the project (Raina, Austen & Timmins, 2012).

CHAPTER 7

Changing Narratives of Change: (Un)intended Consequences of Educational Technology Reform in Argentina

Inés Dussel, Patricia Ferrante, and Julian Sefton-Green

Introduction

Argentina is among the first countries in Latin America, together with Uruguay, to implement a universal program for introducing one computer per child in classrooms. Focusing on secondary schools, it promises to distribute three million netbooks to every student and teacher in public institutions over a three-year period (2010–2012), of which over two million (67%) have been distributed by June 2012.[1] The main rhetoric of the program is centered on social inclusion—hence the name "Conectar Igualdad" (translated as "Connect Equality"). Along with other policies of the Argentinean government in recent years, this initiative has defined a new policy scenario that pivots around egalitarianism and social justice. This chapter uses Conectar Igualdad as a way of reflecting on what it might mean to approach social inclusion through educational technology policies as a wider political action. We are particularly interested in examining how educational technology policies support, hinder, or change school reform and what might be the consequences of these political initiatives on schools, teachers, and young people themselves—almost as unintended consequences of the grander politics that animate their inception. Schools are already the object of explicit political interventions and play a key role in the social imagination as places for inclusion and social change, but we suggest that these "older" and more entrenched narratives are both implicitly and explicitly modified through narratives of technological change.

We thus organize our chapter by first discussing the main political rhetoric that frames Argentinean educational technology policies oriented toward social inclusion. We argue that the state, and not technological companies or global projects, is the central actor at every stage. We then approach the huge challenges that a program as large in scope and scale as Conectar Igualdad poses to everyday life of schools and to pedagogical practices. We then ask what pedagogical changes are required to build up a fully digitized school and how educational technology programs might coexist with policies and practices from other traditions as well as with teachers and principals. Finally, we briefly present the evaluation strategies that are currently measuring the program's impact, describing the first raft of findings, and discuss its effects on schools and teachers through some case studies. Our conclusion argues that Conectar Igualdad gained legitimacy across a wide section of the country through its rhetorical appeal to strongly held beliefs in the state's primary function as an engine of social equality. However, once launched, a process of "vernacularization" (Thomson, Jones & Hall, 2009) took place as all the different social actors involved in the plan (teachers, principals, schools, and young people) found different ways of interpreting the offer made to them though educational technology. We also suggest that in the Argentinean context, Conectar Igualdad represents a form of local social policy almost aside from the wider global arguments about competitiveness and the knowledge economy.

Understanding the Social, Political, and Economic Context behind the Conectar Igualdad Program

Conectar Igualdad was launched in 2010 as a much-trumpeted program to reduce the digital divide and improve public schooling. It received wide media coverage and was heralded as one of the most ambitious educational technology programs in the world, due to its scope and time frame. Not only were three million netbooks to be delivered in three years, but also connectivity, electric wiring, and plug-ins had to be provided for over 13 thousand schools throughout the country. The presidential decree that created the program framed it as part of the recognition of education as a public good and of the personal and social right to a high-quality education. The language was not only one of rights but also of the state's responsibilities. The argument has been made that in light of the enormous social and economic changes brought by digital technology, new measures are demanded from the state. Unlike much global jargon about individualism and liberal freedom, the central statements of the policy refer to the need to level an unequal playing field through the intervention of the state. This state-centeredness is linked to a Latin American left-leaning climate that has renewed the political leadership in most of the region's countries. While the performance of these governments is controversial in many respects, there is no doubt that there has been a remarkable change in political rhetoric and strategies since the 2000s (Weyland, Madrid & Hunter, 2010).

There are some particular events in recent Argentine history that help contextualize this initiative. The economic crisis in 2001—which provoked a plunge of the country's gross industrial product (GIP) and a peak of poverty, indigence, and unemployment rates—was accompanied by a severe political crisis that, despite the initial force of participatory, popular rallies, was succeeded by an increased centralization of the state. The successive Kirchner governments, in office since 2003, developed wide redistributive policies and protectionist economic measures, of which the Universal Allowance per Child, *Asignación Universal por Hijo* (AUH), is a flagship program.[2] These administrations have therefore championed a political rhetoric based on healing social wounds and delivering social justice—perceived as an action whose main protagonist is the state, as the only guarantor with enough power to counterbalance market forces and international agencies such as the International Monetary Fund (IMF) and the World Bank. This period has also seen a rebirth of nationalism and a new sense of national pride based on moral-political grounds ("we could overcome the crisis on our own terms"), human rights policies, advanced civil rights (laws on same-sex marriage and parenthood, legal recognition of transgender identities, and euthanasia) as well as the ambition to develop a new media landscape (with laws limiting the power of monopolistic media and promoting public media). Redistributive policies and state-centeredness have therefore been the two main traits of social and educational policies in the last ten years. Against this background, it is perhaps unsurprising that Conectar Igualdad builds on the acknowledgment of a digital gap to propose a state policy that will bridge this division and reduce social inequalities.

However, Argentina's educational technology policies have to be read along similar experiences in Latin America, where access to digital technology has been a campaign promise and a state policy in several countries (Lagos Céspedes & Silva Quiroz, 2011). Two cases are noteworthy. Uruguay was the first country in South America to implement the One Laptop per Child (OLPC) program—and one of the first countries in the world to scale it up to the whole primary level (Warschauer & Ames, 2010). In December 2006, the Uruguayan government launched the Plan Ceibal—standing for "Conectividad Educativa de Informática Básica para el Aprendizaje en Línea" while also playing on the name of the national flower "ceibo." This policy was seen to place Uruguay at the forefront of the battle for equality, and this concern has been central to the design of the program and of its monitoring (Pérez Burger, 2009). The experience of OLPC in Perú is also framed as part of social-inclusion policies. Smaller in scale than Argentina or Uruguay, the OLPC Peruvian program began in 2007, and focuses on the most disadvantaged areas, predominantly rural and indigenous (Warschauer & Ames, 2010). The program is set to extend to urban areas in the near future. This model has been described as going from the periphery to the center, and in practical terms faces major connectivity challenges due to the isolation of rural areas. Evaluation studies concerned about its impact on academic achievement scores, motivation, and behavior patterns have shown diverging results (Cristiá, Ibarraran, Cueto, Santiago & Severín, 2012), and have largely failed to explore other type of effects on digital literacy skills and children's relation to new media.

Whereas Uruguayan and Peruvian policies have been focused on primary schools,[3] Argentina's educational technology policy has focused on secondary education, targeting all public schools nationwide (in excess of 13,400 secondary schools). Secondary education was seen by the Argentinian government as the optimum site to enhance both social inclusion and educational outcomes. Consisting of between five to six years of schooling and including what in other countries is considered lower and upper secondary, secondary schooling was made compulsory only in 2006,[4] after two decades of consistent growth since the end of the military dictatorship in 1983 (Ministerio de Educación, 2009).

This recent historical situation reflects the sensitive positioning of secondary schools within policies against inequalities, both because of the popular demand to expand universal schooling and also because, notwithstanding its growth, there remain significant disparities in its outcomes. Indeed, secondary education is the stage of Argentinian education with the poorest performance rates of all, characterized by a 38 percent rate of above-age school population, 11 percent dropout rates (a figure that reaches 30% in the last two years of schooling), and low scores in Program for International Student Assessment (PISA) and national tests (UNESCO-IIPE-PNUD, 2009).

Moreover, these figures are tied closely to income inequalities. For example, almost 40 percent of students in public secondary schools were the first in their families to have received a secondary-level education, while 16 percent of teenagers between 13 and 17 years from the 20 percent lowest-earning families are still not enrolled in school. Conversely, the correlation between private schools and the Argentinean middle and upper classes is strong. In the families of the lowest-earning fifth of the population, 91.2 percent of children attend public schools, only 25 percent of the children from the highest-income fifth of the population attend a public school (Rivas, Bazem & Vera, 2010). Thus, it seems evident that any state policy oriented toward public secondary schools stands a good chance of reaching low-income students.

The Conectar Igualdad program therefore includes a strong call to make public schools stronger and more appealing for young people, renewing its pedagogies and bridging out-of-school and school cultures. Unlike Plan Ceibal, which has moved slowly toward pedagogical discussions (Rivoir, 2010; Bañuls, 2011), Conectar Igualdad was conceived as a plan to simultaneously bridge digital inequalities *and* transform schools. As such, where new technologies are seen as coming into schools to update school knowledge, meeting labor-market demands of a digitally literate workforce, and producing digitally competent citizens (Ministerio de Educación, 2011a). One significant point in the rhetoric of the program is that the government has resisted endorsing promises that distributing netbooks will change schools immediately. Instead, Conectar Igualdad has been presented as a measure that pays a past debt to low-income populations, which intends to restore legitimacy, appeal, and social recognition to the public school system, which used to be a source of pride for Argentinean middle classes. As the National Education minister said,

> We don't believe a technological artifact will produce a magic trick in schools or in the classroom…We are not overestimating the situation and saying, "netbooks are here, and the next day Argentinean education changes." Far from it! We make it clear every time we can. (Sileoni, 2012, p. 74)

As such, educational technology policies have been presented as another step in a long-term strategy of improving schools, particularly public schools, as significant learning environments. Netbooks are not seen as substitute teachers or books; access to knowledge and literacy practices is a goal that has to be updated but not abandoned, and that still needs to take place in schools. The official discourse of the program therefore extends a hand to teachers to embrace this renewal and become leaders of the next stage:

> No [successful] policy can be done against the teachers, nor can it be done without the teachers... [Teachers'] positive predisposition is a key issue for the success of this policy, and it obliges the State and it challenges us [to be up to it]. (Sileoni, 2012, p. 75)

The rhetoric surrounding Argentinian education is therefore that access to digital technology will level the playing field, and that this is plausible because it comes as a neutral medium with democratic potentials. The program's advertisements as well as official statements from the president and senior ministry officers—frequently present when the netbooks are distributed—take care to emphasize the fact that everybody is receiving the same machine, and that the reactions of the students are similarly enthusiastic and curious. Interestingly, the notion of play is not mentioned, as it might undermine the legitimacy of "official" academic processes.

Two main ideas can therefore be seen as underpinning the Conectar Igualdad project. On the one hand, the goal of the policy is to universalize access to digital technology throughout the educational system as a way to democratize access to knowledge. On the other hand, the policy is also portrayed as placing the nation-state as the guarantor of access. The rhetoric surrounding the Conectar Igualdad program is therefore replete with references to the central role of the state as the agent responsible for everybody's well-being (cf. Laclau, 2005, for a thorough discussion of this debate). However, it is also important to note that the linkage between technological development and social-inclusion goals is not entirely new in the political discourse in Argentina. Traditionally a primary-goods producer since the nineteenth century, the political planning involving technological development is mostly associated with the model of industrialization for import substitution that took place in the mid-twentieth century. Although military dictatorships throughout the latter half of the twentieth century dismantled this scientific and technological structure, the reestablishment of democratically elected governments in 1983 involved a struggle to recover the country's past strength—despite a succession of economic crises and political instability. Thus, it is only recently that a new technological development program has been agreed (an example of which has been the creation in 2007 of the National Ministry of Science, Technology and Innovation for Productive Purposes). Software development and computer and creative industries are now receiving support via fiscal loans and fellowships for training, and have experienced considerable growth over the past ten years—particularly in the area of small- and medium-sized enterprises (López, 2010). Tellingly, these economic policies have tended to be characterized through discourses relating to global competitiveness—thus contrasting with the educational-policy emphasis on social inclusion.

Making It Happen: Acknowledging the Complexity of the Implementation of Conectar Igualdad as a Large-Scale Program

Considering its scope and range, Conectar Igualdad has surpassed all the other Argentinian policies in the area of education and technology. The past 20 years have seen several national and local projects, most of which focused on equipping IT labs in schools (Dussel & Quevedo, 2010) and introducing "Computing" as school subject in upper-primary and secondary schools. Media education, however, has never received wide support or indeed an officially recognized place in the curriculum. An important initiative during the 2000s was the launch of a national educational portal, Educ.ar, by the National Ministry of Education. This began with a private donation of US$10 million in 2000 from the Varsavsky Foundation but grew considerably in the years that followed to include TV channels and the production of transmedia digital content.

Against this background of partial reform and isolated initiatives, the decision to implement a policy with the scale, costs, and dimensions of Conectar Igualdad has reorganized the field of educational technology within Argentinian education. In particular, one has to consider the concurrent goals of procuring netbooks, establishing connectivity to the schools, providing teacher training for over 400,000 teachers and school principals, and producing educational software. To achieve these goals in less than three years proved in itself an organizational and administrative challenge that was hard to meet—not least in light of the bureaucratic nature of the Ministry of Education. The preferred solution to this issue was to involve several state agencies in the administration of the initiative, thus distributing tasks and responsibilities. The result of this has been that the program is run by multiple agencies with a complex arrangement of responsibilities and division of labor. This has sometimes led to a duplication of responsibilities and a degree of rivalry between these agencies (a phenomenon not new to intersectorial policies, cf. Cunill Grau, 2005). The execution of the initiative has also been assisted by an intergovernmental agency—the Organization of Ibero-American States (OEI)—which has organized the bid offer as well as part of the teacher training. As such, the complexity of the administration of the Conectar Igualdad program has not been a minor fact in its implementation.

The hardware for the netbooks was developed by a consortium of ten international companies, and assembled in Argentina. The resulting netbook device has been designed to run both on Windows and Linux and other free software programs and applications, and includes a wide range of educational software and multimedia tools for producing and recording sound and video. Reportedly, Microsoft granted full license of Windows Office at US$3 per netbook. There are over five thousand educational resources for teachers in the netbook's "desktop" space—mostly produced in previous years by the national educational portal, Educ.ar, and also provided by private publishing houses. Besides the netbook development itself, connectivity has proven to be among the top challenges facing implementation.[5] In Argentina, this necessitated a strong intervention in school infrastructure in the aftermath of years of underinvestment. The ongoing aim of the initiative has been to establish each classroom as a "technological floor" (i.e., establishing adequate plug-ins and electric wiring) that allows the connection of up to 30 netbooks simultaneously. So far, the distribution of netbooks has progressed more quickly than the wiring of schools. According to an external evaluation conducted by independent university researchers, out of an overall population of approximately 13,400 schools, 5,400 schools either have got or will receive internet connectivity in the near future (Ministerio de Educación, 2011b, p. 18). As might be expected, some administrators and school principals have been concerned with the slow (or inadequate) progress of connectivity and wiring. However, teachers and students have found creative ways of dealing with the shortcomings, working offline in classrooms and online at home, as will be shown later.

The Influence of Conectar Igualdad on the Reorganization of Pedagogy

As noted above, Conectar Igualdad has a strong pedagogical appeal to transform schools. As such, important components of the program are teacher training and curriculum policies—both of which have proved difficult to align in terms of time deadlines and resources. Educational systems move at a slower pace than the anxiety of reformers (Viñao Frago, 2002). In the case of the Conectar Igualdad reforms, this has been exacerbated by the fact that there are several agencies running the program, making it difficult to organize a consistent teacher education strategy (e.g., during 2011 there were as many as five agencies offering teacher training).[6] The teacher-training element of Conectar Igualdad has consisted of regional and national meetings with school heads and inspectors to discuss strategies and steps in the adoption of the new technology. These meetings have been complemented by the development of online courses for teachers (run by the OEI and the Ministry of Education), and curriculum materials that develop criteria and examples of teaching units. According to a recent report, over a hundred thousand teachers have received some kind of training, although this training includes self-assisted courses (that is, prepackaged activities) as well as tutored ones.[7] Among them, the OEI offer of an eight-month long digital literacy course with over 150 tutors, has been the most successful and has alone reached over 65 thousand teachers during 2010 and 2011.

Generally speaking, the program documents and materials promote the centrality of teachers for educational change. For instance, the *Guidelines for Classroom Strategies* published by the Ministry of Education (assumed to act as a manual for school principals and teachers for the implementation of the program) proposes as a set of general principles that

> the teacher generates change and gradually incorporates the use of equipment according to her or his goals, training and classroom reality... The teacher will make a progressive use of equipment when s/he feels more familiar with technology, and will increasingly use it in its classroom practices. (Sagol, 2011, p. 13)

These guidelines therefore assume that there will be a gradual movement of teachers from an exploratory moment with computers with or without students, to a regular and then intensive use where computers will comprise the main learning environment. Great care is therefore taken within these documents to stress that there will be an array of levels of involvement carefully including novice and less-trained teachers.

However, this caution and openness has perhaps not been pursued so clearly in relation to pedagogical strategies. The guidelines for pedagogical transformation are notably more celebratory and tend to convey a somewhat simplistic trust in the affordances of digital technology. Also notable is the relative lack of references made to potential conflicts between digital technology use and traditional classroom practices. For example, the Conectar Igualdad guidelines state that in order to benefit from the presence of digital technologies in the classroom, teachers can use digital content (i.e., use the internet as a set of educational resources), digital publishing, social media, multimedia materials, create blogs or projects, promote collaborative works, and/or teach to learn how to manage information (Sagol, 2011, p. 19). Each of these possibilities is illustrated with two-page explanations that provide a synthesis of the media involved but which does not develop substantive discussions of its possibilities and challenges in a classroom setting. For example, the guidance relating to "using social networks," a topic that has received wide discussion around the world (cf. Boyd & Ellison, 2007; Levinson, 2010), stresses only the positive outcomes and marginalizes the possibility of any difficulty:

> A network is a set of interconnected nodes...It is an open and multidirectional structure. It is very likely that students use their networks in the context of leisure or gaming. It is important that the school uses them with the goal of interchanging educational content, problem-solving, decision-making, project development, etc....Networks connect students in new and different ways. It is interesting, for example, that some girls and boys in each form may belong to different networks. These differences will nurture the knowledge and the experience of the classroom. (Sagol, 2011, p. 22)

In contrast to the heated debates that are taking place around the world over the difficulties of accommodating social networks in classroom settings, the official Conectar Igualdad suggestions for classroom use are presented as clear-cut options:

> Use informational platforms to create the networks, instant messaging systems (Twitter), shared documents (Google Docs, YouTube, Delicious), social networks systems (Facebook, MySpace). With digital networks it is possible to replace and improve old communication systems, such as phone chains or bulletin boards. Mounting a networked system of efficient and up-to-date institutional communication helps to generate a sense of community in the group. Networking is advisable in classroom activities that have a certain degree of complexity or require combining heterogeneous tasks. (Sagol, 2011, p. 22)

How these guidelines will help empower teachers, as is their claim, is not clear. Given this level of generality and the thinness of pedagogical advice, it is likely that inexperienced users will not know how to proceed, and experienced ones will find them irrelevant. Nevertheless, the document repeats that "the more autonomy students have in their learning, the more necessary the teacher's role becomes. With their individual equipment, students need a permanent guide, monitoring of its use, and a mediation for information consumption" (Sagol, 2011, p. 15). However, there is almost no discussion in the materials on how to monitor or mediate the relationship with technology. This is left up to each school or even for each teacher to decide.

Understandably, the pedagogical strategies that have been developed in practice to support the implementation of the program at the classroom level and help teachers deal with the complexities of new digital media have moved at a slower pace than the introduction of machines. The production and adoption of new repertoires of practices has taken noticeably more time and encompassed many more dimensions than cabling and computers. The fact that the pedagogical integration of the machines has not been prioritized as a first-order issue during the implementation of the program may well, in the long run, hinder the potential of the program to renew classroom practices and strengthen public schools. A similar observation can be made in relation to the human resources that are likely to be needed to make the program work at the school level. From its inception, Conectar Igualdad has proposed the creation of a new staffing position in schools to take charge of equipment and connectivity. This agent is called "technological referent in school" (Referentes Tecnológicos por Escuela, RTE), and is intended to relieve teachers from the demands of developing technical expertise. However, in practice these positions have proven difficult to fill given Argentina's shortage of technical graduates and (particularly in a time of low unemployment rates) the relatively uncompetitive level of educational salaries. Thus, several school districts have had to divide RTEs between several schools at once, thus compromising their ability to offer schools day-to-day technical support. The question is thus how far these absences and difficulties undermine the aspirations of the program or whether even partial implementation goes some way to producing some kind of change. This is something that will be discussed in the next two sections.

Evaluating the "Success" of Conectar Igualdad

Unlike most public policies in Argentina that come and go without any evaluation (Herrera, 2006), Conectar Igualdad has been the object of several (and sometimes overlapping) studies of its impacts. On the one hand, the program is being evaluated by internal agents at the Ministry of Education, whose methodological design intends to evaluate social inclusion.[8] Conversely, an evaluation being conducted by a group of 11 public universities is focusing on different dimensions such as family use, pedagogical strategies, and literacy practices. While the former evaluation has only made public its design and preliminary baseline findings, the latter evaluation presented a report in October 2011 describing the first stages of the implementation, focusing mainly on teachers' and students' perceptions (Ministerio de Educación, 2011b). Other agencies such as the Presidential Office for Conectar Igualdad and the OEI are also producing evaluations about their own initiatives on teacher training and pilot-schools innovation programs. Given the high financial and political investment in the program, all these reports can be seen as positioned halfway between contributing to its political legitimacy and producing knowledge about a large-scale and ambitious program to bridge the digital gap and transform schools.[9]

One of the findings to emerge to date from these evaluations has been, perhaps not surprisingly, that levels of adoption of computers have been highly variable among teachers. For instance, 68 percent of respondents interviewed in a wide sample of 1,447 teachers said that they used a computer at least once a week. Interestingly, this figure was slightly higher in schools located in areas defined as "emergency scenarios" (i.e., extremely deprived areas)—a finding that has been taken by officials to signal a significant advance with regard to the initiative's social inclusion goals (Ministerio de Educación, 2012, p. 129). This intensity of use certainly appears to be high when compared with the impact of the Uruguayan Plan Ceibal in secondary schools, where 37 percent of the teachers report a similar use (Departamento de Monitoreo y Evaluación del Plan Ceibal, 2011). Indeed, most teachers in the Conectar Igualdad program (81.3%) report that they use the netbooks to plan their lessons and find resources and materials, and 56.3 percent acknowledge changes in the type of activities they are doing in classrooms since the netbooks were introduced into their classrooms (Ministerio de Educación, 2012, pp. 133–136).

Similarly, the teacher-training components of Conectar Igualdad appear to be having an impact, despite not being mandatory. Most of the teachers included in the official evaluation had already taken courses on digital technology use (particularly in terms of the use of Microsoft Office applications). However, the specific training courses on Conectar Igualdad appear to be making an additional difference. For example, preliminary findings suggest that when teachers take training courses they get to know and use the resources provided by the netbooks—most of all the Teacher's Desktop—and produce more complex teaching units. Nearly half of the teachers in one evaluation reported that they are now using the netbooks for evaluation purposes (Ministerio de Educación, 2012, p. 132). Perhaps counterintuitively, this rate drops to 35 percent in the case of younger teachers. According to these figures, teachers are using the netbooks to produce teaching resources of their own. Even though "slideware" presentations are the most popular form of content creation (35.6%), multimedia productions such as videos or edited images are also being produced by over a third of teachers, with 19 percent of the teachers reporting using netbooks for developing whole teaching units (Ministerio de Educación, 2012, pp. 135–137). Of course, digital content might be considered as a "resource" for teaching, but this does not necessarily mean an endorsement of new ways of teaching and learning, or an acknowledgment of new literacies at play.

If teachers' enthusiasm about digital technology can be said to be high, the same goes for students who have been found to be very satisfied to "own" a netbook,[10] which in many cases is the first computer in the family. There seems to be a specific impact in students' motivation, as they feel they are receiving something valuable. For instance, 61.6 percent of students are reporting that Conectar Igualdad has helped to reduce the difference among students who owned a computer previously and those who did not. Similarly, 80.5 percent are reported to consider the program as having helped to reduce educational gaps (Ministerio de Educación, 2011b, p. 47). The majority of students also say that they use the netbook in the classroom for learning purposes (83.5%) (Ministerio de Educación, 2012, p. 140). However, it is not obvious how this translates into other kinds of improvement, and needs much more investigation and follow up.

All of the evaluation reports produced so far concerning Conectar Igualdad have also studied the kind of practices that students are undertaking with the netbooks inside and outside of schools, but the dimensions and instruments taken are not always helpful to understand them. For example, in both reports, using Microsoft Office is listed as the most frequent activity in and out of school as reported by the students (Ministerio de Educación, 2011b, p. 54; Ministerio de Educación, 2012, p. 141). Chatting, gaming, and using social networks are reported as almost nonexistent practices within schools (5%–10%), which seem to be greatly underreported.

Evidence of "Impact" in Terms of Whole-School Change: Contrasting Evidence from Qualitative Research

These preliminary findings certainly celebrate the program, and highlight an understandable sense of ownership by students and a renewed enthusiasm by most teachers. Yet there is still no clear knowledge about whether schools might be improved by the adoption of netbooks, or how teachers are negotiating ownership of the technology in classrooms. Also, important questions remain about whether schools are effectively preparing students better for the contemporary challenges of mediatized societies. There is therefore room for wider considerations of the "impact" of Conectar Igualdad beyond these official evaluations. As such, several interesting case studies have emerged in different districts produced by various organizations—not least school inspectors and teacher educators. These "unofficial" accounts take a wide range of forms—notes, videos, sound recordings, blogs, testimonials, interviews, forums, and sharing meetings—and reflect a more nuanced picture of the state of mobilization and implementation of Conectar Igualdad throughout the Argentinian educational system.

For example, it is clear from these school-level accounts that lack of connectivity is a major obstacle but it would also appear, at least so far, that teachers are finding creative ways to overcome connection-related problems. Official reports from the program state that only 48 percent of schools are equipped with intranet and internet connectivity. In this context, teachers at a majority of schools are still having to either bring digital resources into school on a pen drive and share resources with their students, or otherwise ask them to search for content at home, download it, and bring it to the classroom themselves. Of course, not all students have internet connections at home, but they go to fast-food outlets, coffee shops, and other sources of "public" Wi-Fi. One teacher reported that her students shared knowledge about where to find accessible sources of Wi-Fi, and also are able to "crack" the codes and passwords of some of these locales (such as McDonald's restaurants). This tech-savvy "work around" might not have been written into the implementation of Conectar Igualdad, but is nonetheless being developed through the implementation of the program.

One case—in the process of being researched by Dussel and Ferrante—is a school that constitutes a singular case both because of its recent history and because of its enthusiastic leadership. We therefore take some time to present emerging findings from this study to illustrate the points raised so far in this chapter. The school is located in an upper-class neighborhood in the city of Buenos Aires, and has been known in the past for its students from aristocratic families—for instance, the infamous minister of economy during the last military dictatorship was a graduate. However, in the last 20 years the student population has changed due to middle-class flight to private schools and because of the inclusion of poor urban students in public schools. This school, and particularly the principal and a group of its teachers, is notable for having embraced the netbooks as a means to motivate students' academic work. The perception of the staff is more nuanced than a "yes/no" decision on whether to use netbooks or not. Instead, they acknowledge that this school's problems are multiple and that netbooks will not necessarily solve them all. Nevertheless, the school sees a value in the laptops—not least because they help to put schoolwork in the forefront of students' daily lives, and thereby encourage students to read and write—which they had proved reluctant to do in the previous years.

Staff at this school therefore not only identify gains in student enthusiasm as resulting from laptop use, but also acknowledge a concurrent set of challenges. For example, teachers report that they fear that laptop-based activities are too often banal or frivolous. Staff also report that while laptops might succeed in increasing interest in content, students do not necessarily want to engage with complex texts, whether visual or written. As one Philosophy teacher argues,

> These students lack basic literacy skills because they had such a poor primary school, so when I tell them to read something relatively complex, they complain and refuse to do it. If I could just sit with each of them and read it together line by line, I would probably be able to break this resistance. But given a 20 or 30-student classroom, it becomes almost impossible. But I want to get there. I am not satisfied at them not reading anything and giving up even before they try. We have to find other ways to do it. But I don't know if the netbooks help or hinder this process... I don't think it's got to do with the technology.

This teacher acknowledges that an individualized teaching program—as the technology promises to achieve at some point—is impossible to deliver in a classroom of 30 students. Those who already have the will and the skills to interact with complex texts can more successfully and richly use the tools and possibilities offered by digital platforms. Yet this leaves the teacher confronted with the challenge of working with those students who are not willing or not prepared to do so. What they are left with are "lighter" or poorer uses of the technology. And what might this lessening of ideals imply for the grand rhetoric about social inclusion?

However, this teacher does see that even these students who are not making the most of their netbook use still make some gains. For example, he acknowledges that all students are forced to participate in digital platforms—keeping up with blogs makes their production visible and thus more accountable. The technology certainly makes it easier for teachers to track their writing and give feedback about it. Yet when reading students' blogs, for example, great disparities are evident in the quality of work. Some blogs are complex and sophisticated, and some others have only three to five entries for the whole year—in all school subjects—of one or two sentences each. The staff values the existence of some interventions, however low level. After all, these are students who previously did not participate at all in classrooms. Yet these continual low levels of computer-based work still says a lot about the actualities of how the technology "overcomes" social and cultural inequalities. Of course, this school is relatively rare in its reflexive inclusion of technology, on the one hand taking into account social media and new literacies but on the other hand placing them under broader concerns about enriching students' capabilities and language skills. Technology is not positioned as a "magic bullet" that will solve everything, but more simply as motivation to refocus on academic work and improve teaching and learning, in challenging conditions.

These findings therefore need to be contextualized against the wider research literature on one-to-one netbook and laptop computers. In some published qualitative research reports (Maggio, Lion & Sarlé, 2012; Vacchieri & Castagnino, 2012) and in our own interviews, teachers do acknowledge that there are changes in their classrooms that can be associated with the individualized use of computers. First, they value the possibility of access to a broader, almost infinite, bank of resources. Second, teachers appreciate the inclusion of multimodal texts both as resources and as productions made by students. Teachers report that they are now using videos and pictures to teach curricular content, given that students have the possibility to see or download films or photos and bring them to the classroom. Moreover, some teachers are also starting to promote producing videos or presentations with images, for example, video clips on literary works or presentations on school fieldtrips to museums.

However, in the context of Conectar Igualdad, what these new types of text are, and how to assess their quality, is still unclear. Teachers have confidence that the visual is better in pedagogical terms, because it catches students' attention (Dussel, 2011). In our study, on the one hand, school students are often assumed to be "visual" learners—yet this is taken as a given and not as a product of a rich visual culture that has its specific possibilities. On the other hand, teachers often lack training on how to assess multimodal texts. While most teachers were satisfied with the high levels of participation that these activities produced, they reported some sort of frustration with the lack of depth or poor quality of the texts. The promise of a video clip about a work of literature is understandably appealing. However, in reality the actual content produced by students took different paths and, according to a literature teacher, the dialogue with the piece the students had to read was almost lost. As this teacher reported, "We never got to the reading. They were very motivated to do a video clip but I think it would have been the same to do it without the literary work, they didn't use it." It might be tempting to read these pedagogic gaps as a way of undermining grander ideals, but these data can also be read as the effect of an increased demand on teachers and raising challenges in a way that could eventually lead to deeper change in teaching and learning.

Another (perhaps unintended) consequence of the incorporation of netbooks in the school has been an intensification of teachers' work. Enthusiastic teachers report that their work hours and work load has extended considerably and state that using the netbooks to teach always demands an alternative plan in case of the event of technical difficulties and sometimes the teacher's working day is extended for "educational" reasons. As one social science teacher that we interviewed described,

> We once had a heated debate in classroom about types of States (neoliberal, welfare and so on) and we continued it overnight on our private chat. I was up until 2 am talking with some of them about these issues. This never happened before.

What all these findings tell us about the effects of the Conectar Igualdad program in schools is that they are mediated by many factors and that it is these factors that determine the actual impact of the program. These findings highlight the importance of the institutional conditions in each school. Where teachers and school leadership work in reflexive ways it seems to encourage more "ownership" of the technology and thus opens up interesting uses of technology—including an ability to tackle the challenges of teaching in adverse conditions and to engage in the more complex understandings of netbooks not only as information technology but also as media.

Conclusions

Unlike many netbook projects around the world and many national educational technology initiatives, we suggest that Conectar Igualdad was imagined and presented primarily as a policy that showed what the state could and should do. This has to be understood in the context of Argentina's immediate history and how relatively new forms of governance needed to consolidate their legitimacy. While many national educational technology programs are described in terms of preparing future citizens for economic global competitiveness, this rhetoric—although present—does not dominate Conectar Igualdad. Instead, Argentina has pursued an ambitious delivery plan that was designed to show the authority and capacity of the state. In line with this, we have argued that conceptualizing the netbooks as part of a modernization agenda acted as a way for the Kirchner governments to show the power and place of the state in developing forms of social inclusion. We have tried to show how older and more entrenched narratives about the place of the state in social policy are both implicitly and explicitly modified through narratives of technological change.

This chapter has also advanced the argument that Conectar Igualdad demonstrates a core feature of the practical implementation of many educational technology initiatives in terms of a process of vernacularization when the different actors and institutions begin to own what starts out as top-down directives. In our own ongoing "unofficial" evaluations, we have been particularly interested in how teachers and students have found ways to take on board particular aspects of the program inflecting practice at a local level. It is obvious that this program is uneven and, almost by definition, cannot live up to the very high ideals behind it. Nevertheless, it does seem as if it is making a difference to the authority and capabilities of teachers making different kinds of possible pedagogical relationships with the young people as well as to the agency and independence of students themselves. Like all of these kinds of programs around the world, we suspect that findings in the evaluation will point to questions of timeframes and the challenges of scale—especially in a country as uneven and unequal as Argentina. It could not be any other way. However, it may be that by offering these new opportunities in these ways—even if that was not part of the policy vision—does mean that the broader social goals are in some ways being reached.

Notes

1. See, www.conectarigualdad.gob.ar/.
2. Created by a decree in 2009, AUH delivers a US$180 annual allowance per child to parents who are unemployed or underemployed under the condition of school and primary health care attendance, extending a benefit that was hitherto reserved for those employed on payroll. This allowance, which might still seem scarce, implies an increase of 120 percent of the amount of the financial aid received by poor families up to that moment (Agis, Cañete & Panigo, 2010).
3. Uruguay has started to distribute XOs in secondary schools in 2010, but according to evaluations from 2012 the adoption has been very slow in this stage. Only 11 percent of the teachers reported a frequent classroom use of the netbooks after one of year of the program (cf. Departamento de Monitoreo y Evaluación del Plan Ceibal, 2011).
4. The National Law of Education of 2006 established upper secondary schooling as compulsory. The previous Education Law from 1994 reached only lower secondary schools (grades 7 to 9).
5. Among many others, see the work by JUSTA AFRICA: http://www.comminit.com/en/node/128050.
6. These agencies are, National Ministry of Education, Educ.ar, ANSES/Conectar Igualdad, Provincial Ministry of Education, and the OEI, an intergovernmental agency that has had protagonism in this area, training over 60 thousand teachers since 2010.
7. According to the evaluation report conducted by 11 national universities for the National Ministry of Education, training courses were attended by 472,242 people including principals, supervisors, teachers, families, and students during 2010 and 2011 (Ministerio de Educación, 2011b).
8. This evaluation is taking a stratified sample with over 480 schools located throughout the country, and is following over 8 thousand students and

4 thousand teachers. The first step has been to produce a baseline to compare future developments.

9. Paradoxically, the internal evaluation is perhaps best equipped to result in detailed information on the development of the program, although it is too soon to tell whether it will be able to do so.

10. Students actually sign up to a leasing contract, but they can take the laptop home during their school years and they receive it as a graduation present— a measure designed to improve graduation rates, which are low, particularly among low-income students.

CHAPTER 8

The Ideological Appropriation of Digital Technology in UK Education: Symbolic Violence and the Selling and Buying of the "Transformation Fallacy"

Timothy Rudd

Introduction

The drive to embed digital technology into schools in recent years has been one of the most significant areas of investment in a shifting educational landscape. In many countries, such substantial investment occurred within a period of perceived prosperity and was often uncritically promoted as the "technological fix" within broader modernization agendas, thereby creating new educational markets and diverting energy, resource, and attention from wider and more fundamental structural and organizational issues. In the United Kingdom, the "New" Labour government's "third way" politics represented a clear break from the traditional Left and the symbolic appropriation of new technologies conveyed a wider modernizing intent. Their aim of embedding digital technologies[1] in schools between 1997 and 2010 led to substantial investment, supported by rhetoric heralding the transformation of education, a need to develop pupils' "twenty-first-century learning" skills, and based on an assumption that such action would somehow assure future economic prosperity (DfEE, 1997). Yet despite the significant emphasis, resourcing, growth in interest, activity, and the emergence of various bodies and organizations seeking to promote and embed technology in schools, it could be argued that such bold claims amounted to little more than hyperbole and rhetoric. This chapter critically reflects on these policies and outcomes, arguing that not only did the UK government fail to transform education, but also the outcome and legacy was the accentuation of broader neoliberal frameworks and principles, through the stimulation of an educational technology marketplace, and the embedding and transmission of ideologically informed assumptions about the nature and purpose of education.

Buying and Selling the "Transformation Fallacy"

The precise scale of the overall investment in educational technology in schools is difficult to state with precision. This is partly due to incomplete, unavailable, or conflicting records and estimations, and the complexity of direct and indirect educational technology funding from different sources and the diversity of investment relating to various public and private national and local programs and practices. Nonetheless, it can be confidently stated that investment in the United Kingdom has been significant and unprecedented. Selwyn (2008a), for example, states that between 1997 and 2007, over £5 billion of state funding, in various forms, was directed toward educational technology infrastructure. Year on year from 2001 onward, schools saw significant increases in budgets for educational technology, with UK state schools committing £577 million in 2009 on digital resources and services, excluding curriculum software and digital content costs (BESA, 2010). Other figures show that the resourcing of new technology in secondary schools grew faster than any other part of the public sector, with expenditure exceeding £1 billion in 2008–2009, as programs such as Building Schools for the Future and the Primary Capital Program further increased the spending on educational technology resources (Kable, 2008). There were numerous other associated funding streams such as the School Development Grants, National Digital Infrastructure Grant (£40 million), and the Home Access Grant (£194 million), alongside the funding of associated organizations, and the expenditure undertaken by schools, local authorities, and individuals investing in related private sector products and services.

Even against the backdrop of a global economic downturn, the main funding stream earmarked for schools between 2008–2011—the *Harnessing Technology Grant*—was initially worth £693 million over 3 years (although this was subsequently cut by £100 million by the subsequent coalition government). A recent Forrester report (see Savvas, 2011) suggests that the educational technology marketplace has witnessed a "perfect storm," with competing governments continuing to seek to gain a share of the global marketplace while schools are continuing to invest in order to differentiate their "instructional offers" and to improve their operational efficiencies. Educational technology therefore continues to attract significant amounts of funding, and one would expect there to be a clear evidence base clearly identifying the impacts of this spending, highlighting under what conditions such benefits occurred, and demonstrable evidence of concomitant systemic transformation. However, while the continued and relentless commitment to investment relayed a powerful message emphasizing the transformational potential of educational technology, this appears to be at odds with the realities of how digital technologies have been implemented in schools.

Educational Technology as Symbolic and Ideological Power

Vast expenditure and governmental rhetoric alone cannot adequately explain the clamor to adopt new technologies in schools. In presenting a critical perspective that seeks to look beyond the rhetoric, it is therefore essential to consider the social construction of perceptions and the broader symbolic appropriation of technology in society and in the education system. A number of arguments can therefore be advanced in this respect. For example, new technologies could be seen as being presented as a symbolic representation of progressive change, which can serve to mask underlying ideology, oversimplify cause and effect relationships, and may ultimately represent an advanced form of technological fetishism (Hand & Sandywell, 2002). As a meditational tool developed and applied within broader social, cultural, and political contexts, various symbolic meanings are attributed to educational technology, yet supporting rhetoric often presents technology as ahistorical and apolitical (Noble, 1984). The incorporation of technology into education is often presented and perceived as devoid of ideological intent, and largely portrayed as a necessary modernizing and democratizing tool. However, as knowledge and the instruments for its composition, are situated, constructed, and mediated within the context and structures in which they appear (Leontev, 1978), the representation, incorporation, and use of technologies in schools could be seen as prone to symbolic representation, and subject to prevailing influences of the context, wider fields, and structures into which they are mobilized, classified, subsumed, and recontextualized.

These arguments all clearly relate to the wider political context in the United Kingdom during the 2000s. During their time in office, New Labour's political discourse, evocative of a "new progressivism" (Giddens, 2002) and reflecting a third way politics, was supported by a powerful modernizing rhetoric. Ultimately however, their education policies broadly retained the wider ideologically informed neoliberal accountability frameworks imposed by previous "New Right" influenced Conservative governments. In replicating broad neoliberal frameworks (Stevenson, 2011), and arguably intensifying the mechanisms for doing so, New Labour endorsed standardized curricula and assessment, externally imposed systems of measurement, competition, selection, and the technocratic accountability of schools. The embedding, promotion, and increased consumption of new technologies aligned with a broader political modernization agenda, created a vision of a system in transition, and presented a veneer of transformation (Cuban, 2001). Yet it could be countered that existing market-focused approaches limited the degree, type, ways, and ultimately the purpose for which digital technology was deployed within classrooms. The promotion, incorporation, and institutionalization of digital technology in a highly structured and recurrently patterned social set of arrangement and spaces, therefore was highly unlikely to bring about the proclaimed transformation without concomitant and significant changes to those broader structures, and the "rules" regulating the system and generating practice (Bourdieu & Wacquant, 1992).

Indeed, it may be argued that to some degree the effects that systems and related processes have on individuals are first internalized and then subsequently externalized through the language and practice of actors within the field (Bourdieu, 1977, 1988a). Ultimately, this serves to regulate and limit both real and perceived possibilities, opportunities and choices in relation to undertaking new and innovative action. The accompanying official discourse may be viewed as presenting a techno-romantic and techno-determinist vision (Benyon & Mackay, 1989, 1993; Goodson & Mangan, 1996) in overemphasizing the transformative and benign effects of educational technology as an autonomous force, while simultaneously overlooking, or understating the significant structuring and mediating effects of neoliberal orientated system requirements. From this perspective, politically motivated claims regarding "transformation" were both fundamentally flawed and prone to failure, without wider and fundamental systemic changes.

"Edubusiness," Ideology, and the Social Construction of a "Logic of Practice"

As well as endorsing market principles and related monitoring and control mechanisms, New Labour's tenure saw the extension of the privatization agenda (Benn, 2011), supporting the increasingly pervasive influence of "Edubusiness" in the schools sector (Ball, 2007). At the forefront of New

Labour's educational technology policies were overt aims regarding the need to enhance economic competitiveness (Selwyn, 2008b) with the de facto assumption that providing industry with appropriately skilled employees is one of the key functions of education. In so doing, it may be argued that a further ideological shift subverting the purpose of schools, and the primacy of educational technologies within them, toward the needs of industry. The role of business and markets in education was consistently and uncritically presented as a positive force, yet the role this played in the social construction of educational technology tended to be largely overlooked.

This symbolic and ideological repositioning was further compounded through the stimulation of supply-side markets, actively promoted and stimulated by the government, which served to create and drive demand by schools, thereby influencing patterns of educational technology product and service consumption. The underlying assumption in much of the accompanying rhetoric implied that the stimulation of an educational technology marketplace would automatically create the dynamism and innovation necessary to drive systemic transformation. However, such assumptions seem somewhat misplaced when the prime motive for private enterprise is to generate profit by providing viable and "saleable" products resonating with the newly created "needs" of "consumers." Ironically, many of the state-stimulated and endorsed educational technology firms orientated themselves toward designing technology "solutions" to enhance efficiencies in relation to existing structures and related performance and accountability frameworks. This often—perhaps inadvertently—served to reinforce perceptual barriers to more radical innovation, thus reducing the potential for system-wide transformation. In other instances, existing products intended for other sectors were adapted quickly for sale into the burgeoning schools marketplace with scant regard for pedagogy or broader cognitive or learning processes.

Given such bold state endorsement, the UK educational technology marketplace has understandably continued to be proactive in selling digital technology to schools perpetuating the perception representative of "the future," attempting to market products and services as a necessary prerequisite for any forward-thinking or progressive school. Yet, this is imbued with both conscious and subconscious messages regarding the nature of society and the role of education within it. Unproblematic visions of an educational future transformed by digital technology therefore, may not only be misleading and based on presumption and conjecture, but can also mask underlying ideology that affects action and practice (Benyon & Mackay, 1989). The ideological subtext is often subtly yet inextricably associated with global capital and benefits to capitalism (Waller, 2007). However, the myth of the neutrality of technology was perpetuated despite the design, use, and application of technology being a product of historical, social, cultural, and political contextualization, and mediated through the fields into which it is supplied, applied, and regulated by the dominant practices that occur within them.

A further issue that needs critical reflection was the presumption that industry would act magnanimously, unselfishly, and in an informed manner in assuring profits in a competitive marketplace. The reality is that the educational landscape, however, among the many worthy digital resources and related services (see Luckin et al., 2012), is still littered with inappropriate products, ineffective guidance, unsuitable content, and strewn with meaningless service agreements. The rhetoric of transformation perpetuated by the government, other agencies and bodies, and vehemently promoted by the educational technology industry seeking to capture new markets, may have contributed to the anxiety of local authorities, schools, and teachers, fearful of being left behind by technologies promise in a high stakes and profoundly competitive schools marketplace. It can therefore be argued that substantial investment and vested industry influence served to relay inculcating messages regarding transformational power of technology and manufactured a perceived need for school investment. Yet, arguably the reality was to create a "hidden curriculum," governing not only what ought to be bought and taught, but also how and why this should be undertaken, thus ideologically influencing the prevailing "logic of practice" (Bourdieu, 1980).

Manufacturing Consent: Financial Investment and Authoritative Discourse

From the outset, numerous official publications pronounced New Labour's third way commitment to "modernizing education for the next century" (DfEE, 1997), which clearly prioritized a vision of education that was key to, "helping (our) businesses to compete and giving opportunities to all" (Tony Blair, ibid.), underlining a wider political persuasion to appease and actively involve industry in the process. The authoritative and symbolic nature of such discourse, on the one hand, attempts to portray educational technology developments as neutral, inevitable, and beneficial for all, but on the other, foregrounds the influence of vested industry interests in policy formulation. In the same publication Blair stated,

Last year, I asked Dennis Stevenson, chairman of Pearson, to conduct an independent investigation into the potential of information and communications technology in schools. His report identified two main problems—the need to train teachers and to create a market for high-quality British educational software. (ibid.)

It is perhaps unsurprising that an "independent" report by the chairman of one of the world's leading education publishing companies, might conclude that there was a need for the government to create and stimulate a market for educational software—a market in which they subsequently became one of the largest suppliers of multimedia tools, testing programs, and a range of related digital learning content. Moreover, it was also stated that the investment in a "National Grid for Learning" would be a tool to address the "problem" of growing the size of the market for software and that the government would "give schools some 'seed corn' funding to buy those services... pioneering this market at home... to create markets for our companies abroad... We believe this strategy will be good for our children and our companies" (ibid.). The official justificatory discourse espoused the need to "skill" pupils for the future, yet presented generalized claims based on broad assumptions and ill-defined notions, lacking in detail with regard to the nature or purpose of such skills beyond the spurious training and vocational needs related to a projected economic future: "Technology has revolutionized the way we work and is now set to transform education. Children cannot be effective in tomorrow's world if they are trained in yesterday's skills (ibid.).

The underlying assumption here is that such skills, or rather some indistinct form of technological literacy, would be crucial for employability and that without such skills learners would be at a disadvantage in the employment market. This can be seen as symbolically inferring an ideologically informed view regarding the primacy of education and its subservience to the needs of the economy. The authoritative tone presented in official discourse also tended to present both an oversimplified view of the future that belied the implicit vagueness in detail, foregrounding technology skills, yet failing to account for the pace or potential of technological developments. The presumption that digital technologies used by children in school, and the purposes for which they use it in the school context, will have any direct or significant relationship to the digital tools used in a complex and diverse world of employment currently, let alone the future, is tentative at best. Of course, the underlying ideological intent of such proclamations may have been as much about creating general technological dispositions fostering consumption patterns and stimulating demand.

Yet in presenting such a view, an overly simplistic determinist argument was constructed that assigned an unwarranted degree of agency to the technology itself, made blind leaps of faith, and masked wider realities and ideology. Such determinist views, constructed within dominant discourse, seldom accentuated or foregrounded consideration regarding human or environmental welfare, or issues of equality, but tended instead to be imbued with implications derived from and promoting the maintenance of existing power relations (Apple, 2004). Skills for the future, functional technology and digital literacy skills, as with technology itself, will constantly evolve, yet how they do so will continue to be shaped by the cultural, social, and political contexts in which they are appropriated and through which they derive meaning. The transmission of values and emphasis on the importance of schooling to support national economic competitiveness, presents a flawed human capital argument (Blaug, 1987), emphasizing the surplus value potentially created in "skilling up" or training learners for the economy, further embedding a hidden curriculum transmitting ideological values, serving to reinforce the role of schools as sites for cultural and social reproduction (Bourdieu & Passeron, 1977).

Thus, while it may be assumed that given the governmental emphasis and investment in the promotion of educational technology, the wealth of evidence and research in the field resoundingly justified expenditure and expansion. However, on critical reflection, it can be seen that the evidence base is far less convincing than the one perpetuated and not as compelling as might be perceived or expected. This too, perhaps, represents a process of social construction influenced and shaped by ideology and vested interests. The next section therefore goes on to consider the social construction of this evidence base.

The Social Construction of Evidence and the Transformation Fallacy

Official publications (see e.g., DfES, 2003; Becta, 2009b) promoted the advancement of educational technology to schools, and numerous "evidence informed" publications and related arguments were put forward suggesting that such investment would lead to learning gains and school improvement. While it is not possible to conduct an in-depth analysis here, even a cursory critical reexamination of some of the key official studies, reports, and centrally funded and commissioned research, reveals that the evidence base for justifying investment was far from convincing. For example, one of the largest early UK studies, prior to New Labour's tenure, the "ImpaCT Report" (Watson, 1993), concluded that computers could have a positive impact on learning but that findings were inconsistent and varied. Yet, the broader conclusion and interpretation in subsequent publications suggested that as teachers, pupils, and schools became more familiar with technology, greater positive impacts would automatically ensue. This not only down-played structural limitations and changes in technology over time, but also recontextualized findings in line with a broader political modernization agenda. Interestingly, the report's author later pointed out that too much emphasis was being placed on the "actuality of the new" in the field and that "the rhetoric for change has been too associated with the symbolic function of technology in society" (Watson, 2001, p. 264).

The subsequent "ImpaCT2 Study" (Harrison et al., 2002) occurring after significant investment, was one of numerous projects commissioned by the Department for Education and Skills and managed by its adjunct British Educational Communications and Technology Agency (Becta), aimed at evaluating the progress of the government's "ICT in Schools" program. It was one of the most comprehensive investigations into the impact of digital technologies on educational attainment, involving 60 English schools. It concluded that educational technologies had, "shown to be positively associated with improvement in subject-based learning in several areas [and that its] contribution was statistically significant though not large." While this suggests there was evidence of impacts against formal subject and attainment areas, the findings may be interpreted in different ways. Again, the assumption was made that increasing familiarity over time would lead to greater learning impacts. Alternatively however, one could also question the extent of any benefits given the investment and emphasis, and moreover, whether any learning or impact gains would be sustained, increased, or become negligible over the longer term. The various types of technologies and precise ways in which they were utilized, certainly brings into question later uncritical government proclamations about impacts, which tended to focus largely on the positive findings.

A further large government project exploring the impacts of technology on learning, teaching, and school organization was the "ICT Test Bed project" (Somekh, 2007). This £35 million government-funded project sought to "saturate" 31 institutions (28 schools and 3 colleges) with new technologies and provide associated professional development. It was also supported by a discrete ICT Test Bed implementation team, as well as an independent external research team. Perhaps unsurprisingly, with an overall expenditure-to-institution ratio in excess of £1 million, the results suggested that technology had had a positive impact on attainment resulting in learning gains "beyond expectation," although more so in primary than secondary schools, and not always consistently or comprehensively. Again, this evidence was used to inform future policy direction and justify both prior and future expenditure in subsequent promotional literature. However, there are some rather obvious issues that need taking into account. First, the resource and expenditure was neither replicable at the national level, nor transferable at school level, and therefore it might be reasonable to assume that attempts to do so would not yield the same reported impacts elsewhere. Furthermore, given the significant resourcing and intense professional development focus, it would have been more surprising if no impacts had been found. Moreover, it is quite plausible that similar, if not greater, impacts had arisen through other "nontechnology" focused interventions if funded to the same value.

Despite what was regularly presented in wider official government publications and the promotional materials of vested interest groups and private providers, the broad and diverse findings relating to the educational benefits of technology remained relatively unclear and clouded by complexity, as even alluded to by Becta itself in one of its reviews of the research publications:

> Over the last few years, independent studies have analyzed the relationship between technology and learning outcomes for school-age learners. These have included interactive whiteboard evaluation studies in primary and secondary schools, the ICT Test Bed evaluation, and the 2002 ImpaCT2 study. The relationship is not a simple one. Time taken to embed the use of technology, school-level planning and learners' skills and models of learning are all important in mediating the impact of technology on outcomes. (Becta, 2009b, p. 24)

This contrasts somewhat to proclamations made the previous year in Becta's (2008c) *Harnessing Technology, Next Generation Learning* strategy. Not only was it stated boldly that all learners need the chance to use technology to support their learning, but also this strategy further positioned the impacts of digital technology use to bring about greater productivity, prosperity, fulfillment, stronger communities, and even a fairer society. The recent global economic downturn, subsequent political austerity measures, and record levels of youth unemployment, to name but a few unaccountable variables, now raise questions as to how overstated such claims are and how they attribute unwarranted democratizing power to technology. Such hyperbolic claims, camouflaged by the use of spurious "evidence" used out of context were commonplace. For example, questionable extrapolation of results based on an interaction between two (of many others overlooked) variables in research undertaken by the Institute of Fiscal Studies (Chowdry, Crawford & Goodman, 2009) was used on government and government agency websites and in other published materials, and subsequently quoted in various other presentations and speeches to the wider educational technology community by government bodies and industry alike. The claim, presented as fact and devoid of contextual information, was that "research shows pupils could improve by two grades at GCSE with a computer at home" (cited in Nutt, 2010). This sort of cross promotion by official bodies overstating the presumed positive effects of using digital technologies was common practice:

> Ofsted recently commented on the benefits gained by learners through using ICT. It concluded that technology was contributing positively to the personal development and future economic well-being of pupils and students (Becta, 2009b).

The commercial influence on government policies was also particularly notable in the decision to heavily invest in interactive whiteboards (IWBs) and associated training, with initial funding on IWB expansion programs alone totaling £50 million (Hall & Higgins, 2005), with subsequent funds and resources being spent on associated training, consultancy, and resources, and the stimulation of private industry in the marketplace. At the time of the "rollouts," there was little clear evidence to suggest IWBs would lead to learning gains. Subsequent research undertaken after IWBs were placed into classrooms suggests the evidence remained far from consistent. It highlights that benefits were mediated by numerous factors, including application, use, training, software, time and curriculum fit, pedagogical aims and practices, and so forth (see e.g., Higgins, Beauchamp & Miller, 2007; Moss et al., 2007; Rudd, 2007), and is often seen as supportive of current rather than transformational practice (Banjali, Cranmer & Perrotta, 2010). Such broader findings again tended to be overlooked in favor of the less than prolific positive evidence. Commercial interests in the field also not only emphasized positive impacts through promotional materials, but were also actively involved in commissioning and conducting "research" to influence the schools market. Allied to this the burgeoning educational technology marketplace also created new opportunities for writers, researchers, and a cadre of consultants to position themselves in a "profit and loss" marketplace, convincing both policymakers and practitioners alike of the transformational power of technology (Nutt, 2010).

This is not to say that criticism cannot be also made of the supposedly independent research community of university-based academics. The wider research literature produced by UK academics during the 2000s should also be considered in relation to the perpetuation of largely uncritical acceptance of the progress of new technologies in education. While the majority of research could be considered to be robust, accurate, and relevant, much of it, understandably, focused specifically on what was currently happening in the classroom setting with emphasis on individuals and schools rather than the broader social context and structures mediating the appropriation of technology. Many academics focused on pilot and intervention projects, new and emerging tools, and resources and best practices, which gravitated toward optimistic representations and possibilities rather than the more mundane realities of day-to-day practice (Selwyn, 2011c). A significant other body of research sought to understand the effects of new technologies in relation to their impacts in relation to cognitive development and processing, learning technology mediated classroom experience, and the process of technology design and use. While these are vital areas for research, the growth in interest in such areas meant fewer critical and macro analyses were undertaken.

Others have questioned the validity and robustness of the evidence regarding the impact of educational technologies, suggesting much evidence contains bias, and that positive findings are also overstated in subsequent reporting and presentation to wider audiences (see e.g., Trucano, 2005) and that vast expenditure has lead to only negligible or unsustainable gains (Pflaum, 2004). Moreover, it is not uncommon to hear claims of the broad, or potential impacts of digital technology, as if it were a singular entity rather than the diverse set of tools used in numerous ways, for various purposes. Furthermore, there is often a relative lack of detailed analysis regarding the specific affordances of technology and how it interacts with pedagogical and classroom practices to improve learning. Much of the evidence recited selectively by the UK government gave little real detail regarding the context or conditions under which any such impacts occur, or indeed why they occur, from a broader educational and pedagogical perspective.

As well questioning why the use of inconclusive evidence regarding the relationship between the use of digital technology and educational benefits, we must also consider the nature and intention of many of the evaluations. It may be argued that the larger government-funded research studies and reports, such as those outlined above, seeking to evaluate the impact of digital technologies in education were often actually post hoc evaluations of the government's prior expenditure and investment in the field. The influence of the funder, their vested interest in relation to their commitments to promoting technology and the aim of stimulating an educational technology marketplace, need to be thoroughly considered in relation to potential influence exerted in shaping the focus, design, and more importantly, the subsequent reporting and re-presentation of findings through other mechanisms and media.

As such, it is perhaps understandable that while evidence of positive findings are often found and foregrounded, other evidence had a tendency to be somewhat overlooked. Evidence in the field reporting issues such as setup times, the impact of failing technology, ineffective teaching with technology, pupil distraction from task, downtime, cost of upgrades and refurbishment, and so forth, tended to be disregarded as issues that would be resolved over time. Moreover, claims regarding benefits tended to downplay the possibilities of interactions due to the numerous other variables that could come into play, the possibility that impacts are short term, or unsustainable due to cost and changing technology. They also surprisingly, given the scale of investment and the excitement technology can cause, tended to disregard whether results were actually the result of some sort of "Hawthorne Effect" caused by specific investigation, or whether outcomes were due to increase in human and technological resources, or related clarity over learning aims as part of the project process. Moreover, in an educational landscape littered with numerous initiatives, policies, and programs, and with numerous individual and institutional factors and variables in operation simultaneously, it is hard to isolate the impacts on learning and attainment scores and attribute them solely to the use of technology within the educational context.

Indeed, the broader literature focusing specifically on barriers to the "effective" use of digital technology identifies numerous real and perceived intrinsic and extrinsic barriers to "innovative" practice in schools, occurring at both micro (teacher/classroom), and the meso (organizational/institutional) levels (see Becta, 2004). However, macrolevel, or educational system-level barriers also have to be accounted for in relation to the effective use of educational computing (Balanskat, Blamire & Kefala, 2006) and in mediating more innovative learning experiences that better harness the affordances of the technology for deeper, collaborative, and creative modes of learning. For example, Banjali et al. (2010) drew on expert perspectives on technology use in European schools who identified that the use of digital technology in schools is not necessarily innovative and is often used as little more than "an up-to-date pen and pencil method." They state that tools such as whiteboards are still dominated by "frontal" teaching methods, often failing to exploit the interactive potential fully into learning experiences and teaching strategies, thus negating potential for more collaborative and creative skills development. They further highlight however, the need to tackle educational innovation in a holistic manner, as changes in practice and curricula can be undermined if they are not matched by similar changes in imposed testing, targets, assessment, measurement, and accountability requirements. In short, structural regulatory frameworks present system-level obstacles to innovation, the development of broader skills and more engaging pedagogical practices, thus negating transformational possibilities. Similarly, prescriptive and content-heavy curricula and rigid assessment methods emphasizing knowledge acquisition and factual recollection, or perhaps regurgitation, results in compartmentalized and decontextualized learning and hierarchical relationships, with a pervasive ethos of control that promotes conformity rather than diversity that stifles creativity.

Balanskat et al. (2006) undertook a meta-level analysis based on evidence from 17 "impact" studies and surveys carried out at national, European, and international levels. They noted that the rigid structure of the education systems impeded the effective integration of digital technologies into learning:

> Sometimes education systems work against ICT impact and even if educators are not ICT-resistant, in some cases the system under which they work is. For example, in UK, national tests are not made for ICT rich schools. Studies such as the Test Bed study give some valuable results concerning the factors that impede the effective use of investments in ICT. As it was shown in the study investments in ICT are not able to have an impact they should have in secondary schools within the present education system...For example, in the ImpaCT2 study some teachers explain that very little use of ICT was made in Key Stage 3 English, because of the need to prepare for the public examinations...Indeed, existing assessment and evaluation methods primarily focus on content and neglect social and other abilities of learners. Competencies such as problem solving, presenting material in novel ways, collaboration or creativeness are only to a limited degree covered in national exams. Students receive no credit for these new competencies they have developed. (Balanskat et al., 2006, pp. 52–53)

Across the research literature there is a wealth of similar overlooked evidence that calls into question the systemic limitations that seriously reduce possibilities for transformation. Innovations, such as new technologies, are defined by and shaped by the context and conditions in which they are used. At a micro- and meso-level context, history, preferred pedagogical practice, and so forth, will all mediate how technology is integrated and used in learning and teaching. However, within the context of a rigid and highly structured field, any great degree of diversity and dynamism is likely to be severely mediated and refracted in line with underlying regulatory logic. Across a rigid, accountable, and prescriptive system therefore, the tendency has been for new technologies to be appropriated to improve efficacy and effectiveness of the current system, rather than to transform, as new technologies lack the agency often attributed to them in official discourse.

Educational Technology, Performativity, and Isomorphism

In the later stages of the tenure of the UK New Labour government, emphasis seemingly shifted toward addressing school and practitioner "weaknesses" in relation to the use of technology, identifying individual and institutional factors mitigating effective use. In an attempt to gain more leverage and influence, and generate further interest and impact, the attention of Becta, the Department for Education and Skills, and a growing number of other agencies, organizations, and private companies shifted to the development of further frameworks and blueprints for effective or best practice. Various matrices, metrics, review frameworks, educational technology "marks" and "accreditation" routes were developed with the aim of fostering somewhat abstract "e-confident systems," "e-mature schools," and "e-confident teachers and learners" (see e.g., NCSL, 2004; Becta, 2006b, 2008c; NAACE, 2008).

This shift in strategic emphasis seemingly intensified the perceived need for investment and use of digital technologies in schools and further valorized and inflated the "currency" of digital technology in the educational marketplace. Such tools identified "what good use of ICT" looked like and provided various benchmarks and action plans for improvement. Significant resources were spent focusing on system efficacy, monitoring, and "performativity" measures as a means for raising standards and competition within existing frameworks, simultaneously identifying areas for infrastructure, content, and practice development (c.f. Ball, 2000). In so doing, greater emphasis was placed on to the role of teachers and institutions in taking subsequent action, resulting in further investment and demand, yet effectively serving to recontextualize technology as a tool for maintaining and increasing efficiencies within the system. As Fisher (2006) argues, while discourse surrounding educational technology espoused system-wide transformation, the reality was often seeing "the same thing done differently" (p. 293), while Hammond, Younie, Woollard, Cartwright, and Benzie (2009) suggest that while there was a significant increase in resources and a whole range of innovations in policy and practice, the central control over the organization of schools, curriculum, and measurement mechanisms remained largely unchanged.

In this sense, rather than promote the transformative practice consistently alluded to, the emphasis in a controlled marketized system resulted arguably in significant "isomorphism" (DiMaggio & Powell, 1983), with schools imitating one another, or developing independently but under broadly similar conditions with set parameters and guidelines, thereby embedding technology in broadly similar ways. Broader frameworks set out (coercive) parameters, and official models of best practice and frameworks of so-called e-maturity resulted in schools responding to uncertainty by imitating other institutions and organizational configurations in the belief that this would be beneficial. Such beliefs also provided the orientation for significant normative isomorphism of the teaching profession, persuaded by the constructed and perceived need to incorporate new technologies into practice, and predicated on and regulated by formative systemic and organization orientation.

Educational Technology, Ideology, and Symbolic Power

The proliferation of new technologies in education during this time can therefore be seen as triggering a burgeoning concomitant network of organizations, agencies, industries, and interest groups that either by design, or by default, overemphasized the impacts and importance of technology. Consistent government messages, reinforced by vested interests in the technology industry and among wider groups and media, perpetuated a largely consensual discourse supporting the seemingly inexorable and largely positive influence of technology in society and education, while propagating specific projections regarding the future of learning, society, and the importance of economic competitiveness. Such discourse was persistently reiterated until it entered common parlance as taken for granted assumptions and gaining general acceptance. As Young (1984) asserts, while we should not look to create "antitechnology" arguments per se, there is a much greater need to be more critical and highlight how technicist approaches become embedded in conventional "wisdom":

> Perhaps the most effective means of ensuring public co-operation is the rapid institutionalization of "computer literacy" through the premature installation of new requirements for schooling and jobs, which literally forces the population to accept a new set of dubious realities. (Noble, 1984, p. 609)

Noble (1984) argued that vocational and market ideologies surrounding computer literacy (such as those promoted and exacerbated under New Labour), which are linked to employability and economic competitiveness, are used to validate the imposition of educational technology policies, virtually free from critical analysis. Others have long espoused the need for more critical analysis, as there is a tendency that the ideology of computer literacy benefits vested industry interests but is presented as commonsense and beneficial for all equally (Goodson, 1992; Goodson & Mangan, 1996). As

Selwyn (1997) suggests, educational technologies may offer many potential educational advantages, yet they are often viewed with excessive optimism. Uncritical arguments presented in relation to the future employability and the nature of the future of work and society offered a case for the expansion of digital technology in education based on the seeming plausibility derived from accordance with broadly one dominant vision of the future of productivity. Perhaps more importantly, in doing so, it also served to further subvert the perceived purpose of education to the needs of the economy and private accumulation. The progressive and transformational message surrounding new technology has been propagated and has become embedded in our cultural psyche, constructing educational technology as a necessity not only in terms of learning potential but also in relation to the skills needed by industry. Drawing on Noble's (1998) prior assertions, it may be argued that the heralded technological transformation has merely provided a "disarming disguise" for the further commercialization of education through the commodification of learning (Cuban, 2001) and the stimulation and growing influence of the private sector producing marketable products for sale to help schools and pupils compete in a neoliberal framework of accountability, measurement, and control.

Thus, we cannot ignore the role educational technology policies have played in serving to intensify schools as sites for social and cultural reproduction through the appropriation of neoliberal language, ideology, rules, and logic (Bourdieu, 1993). In subverting further the purpose of education to the needs of capitalist accumulation and wealth generation through the creation of a marketplace producing commercially viable products to service, intensify, and further reify existing frameworks and practices informed by human and intellectual capital inference and technicist delivery models, symbolic violence through pedagogic action has also occurred (Bourdieu & Passeron, 1990). The authoritative discourse within a framework of disciplinary power in a highly structured and regulated field increasingly influenced by market forces, resulted not in transformation, but an intensification and digitization of existing practice through the construction of idealized forms of conduct and practice. The discourse and language of transformation and modernization were subsumed within a regulatory neoliberal framework, which ultimately meant that the more innovative possibilities for technological support for learning were lost through incorporation into existing everyday practices normalized by the dominant logic operating and exerting control over the field of education (Foucault, 1977).

In this sense, we must consider educational technology as playing a role in the inculcation in the dominant orthodoxy and hegemonic representation of worldviews and discourse, concealing alternatives and reproducing unequal power relations (Hoffman, 2004). The neoliberal consensus pervading language and structures in education, and in broader commonsense interpretations of the world (Bourdieu, 1998a, 1998b; Harvey, 2005), positioned and mediated the potential of new technologies predominantly as a "techno-centric fix" (Kvasny, 2006), or force for modernizing a competitive system, serving to make more efficient existing modes of accumulation, while making unfounded claims suggesting improved access to technology would somehow lead to improved life chances and mobility for all. It was not meaningfully constructed as a liberating force for learning, or as a tool to enhance broader social justice and equality, or as a tool to challenge systemic and patterned dimensions of disadvantage and inequity.

Conclusions

The embedding and proliferation of technology in UK education during the 2000s was clearly located within New Labour's "modernization" agenda, which justified its emphasis in relation to a seemingly inevitable and irreversible process of change, partly determined by a perceived unstoppable and unprecedented process of technological advancement. Presented as a fait accompli, it received far less critical attention, and furthermore, it drew on broader neoliberal vocabulary and associated "newspeak" (Bourdieu & Wacquant 2010) that has come to pervade our language, discourse, and media, and which has diffused as a new "planetary vulgate." Drawing on such a perspective, there was an emphasis in accompanying discourse around terms such as "globalization," "flexibility," "governance," "employability," and "new economy," while terms such as "capitalism," "class," "exploitation," "domination," and "inequality" are conspicuous by their relative absence, deemed largely irrelevant in political and public discourse.

This persistent drive can therefore be seen as representing a new form of imperialism whose effects are even more powerful because it is not only directly promoted through partisans of the neoliberal revolution whose intention, under the banner of modernization and transformation, is to brush aside, "the social and economic conquests of a century of social struggles" (ibid. p. 1) but is also perpetuated by the "cultural producers," such as researchers, writers, practitioners, evangelists, and activists, who may perhaps still perceive themselves as progressives or opposed to neoliberal agendas. From this perspective, the UK educational technology drive could be seen as representing a form of symbolic violence (Bourdieu, 1990) in that it relies on a relationship of constrained communication in order to dehistoricize, universalize, and create misrecognition (Bourdieu, 1998), positioning individuals as responsible for their own position in the world, while masking the role of ideology and societal structures (Bourdieu et al., 2000) in the creation of a priori judgments. In this sense, the fallacy of educational transformation has been constructed and perpetuated as neutral process, yet has progressed through truncated dialogue serving to mask broader ideological bases and consequences, with those opposing or resisting its inexorable proliferation into education often perceived as luddites, antitechnology, laggards, resistant, unnecessarily skeptical, or out of touch.

Despite how these conclusions may be read, given the interests and investment in the field, this is not an antitechnology argument. Rather it is one that offers a critical reflection of the wider macro conditions and ideological influences at play that have regulated and generated practice and action. The ideological influences not only negated the potential for systemic transformation and more innovative learning and teaching but also have served to structure and further embed unequal power relations within the education system. Understandably, educational technology offers great cause for optimism as it represents a set of potentially powerful tools that can assist and enrich learning and teaching in previously unimaginable ways, and the field offers many striking examples of innovative, creative, engaging, meaningful, and collaborative learning facilitated by the affordances of technology. Yet perhaps the most striking and beneficial examples often appear on the margins, when technology is specifically applied to address particular or special educational needs, or where they occur outside or on the edges of accountability structures and systems of stringent measurement and control. While the design, appropriation, and use of technology is not automatically political or ideological, the fields into which it is incorporated are, to a greater or lesser degree. The structuring effects of such highly prescriptive and unequal fields, systems, and structures, which reflect the needs of particular interests over others, serve to imbue practice and use with an often-unconscious set of actions that relay and reify broader political and ideological intent, propagating symbolic violence and reproducing underlying power relationships.

Reiterating one of the themes in the opening chapter of this book, a key issue there is what is *not* being said throughout this social construction of educational technology—that is, the gaps, silences, and omissions. Because of the ideological orientation and inequalities inherent within the system, issues of equality, social justice, and democracy are not of central concern but are peripheral, often "add ons," afterthoughts, or constructed around institutional or individual deficit models, based on truncated language or skewed evidence that mask broader principles and alternative ideals. Of course, educational technology *could* be constructed differently—that is as a liberating tool for equality, empowerment, social democracy, and wider socially good purposes. Yet "social good" is framed by hegemonic discourse in relation to economic competitiveness, accumulation, competition, and mobility. Technology is unlikely to bring about systemic educational transformation within a system of unequal power relationships. Perhaps it has more transformational potential outside and on the margins of the restrictive confines of regulatory systems, through acts of resistance that challenge dominant orthodoxy and the tyranny of the market.

Note

1. Over the specific period, these tended to be referred to as either information technologies (IT), or information and communication technologies (ICT).

PART IV

The Politics of Education and Technology: Extending beyond "the Digital"

CHAPTER 9

Mobile Technologies and an Ethical Digital Literacy in the Face of Empire

Richard Hall

Introduction

Mobile and wireless technologies are often described in terms of the efficiency and productivity they enable both inside and outside of formal educational settings, through their provision of "flexible and timely access to learning resources, instantaneous communication, portability, active learning experiences and the empowerment and engagement of learners, particularly those in dispersed communities" (JISC, 2012, n.p.). Given the increasingly proscribed research and funding agenda within contemporary tertiary and higher education, pedagogic case studies for the learning that is enabled through personal, mobile tools increasingly focus upon value-for-money and cost reductions in educational institutions, or on the impact of personalized outcomes for learners and teachers. Such cases rarely critique mobile technologies beyond the pedagogies deployed, any limiting technical issues, and the spaces and places in which they are deployed.

Such a limited field of analysis, although pedagogically driven and often focused upon participation or (re-)engagement in learning, risks describing mobile technologies as socioculturally neutral, against their absorption within or description of social relationships and networks of power (Feenberg, 1999). Yet it is possible to analyze how those mobile technologies and their production/consumption form nodes in an integrated network of restrictive "polyarchic" governance, which closely delineates what can be critiqued or discussed inside the structures of liberal democracy that are themselves insinuated inside the logic of capitalism. This chapter starts from the premise that mobile technologies and mobility emerge from within the logic of capitalist work and the realities of labor inside capitalism. It is the nature of this relationship between mobile technologies, mobile learning, and capitalist work that therefore frames this chapter.

In developing a more critical view of mobile learning there are two areas that might be developed, first, *against* pedagogies of consumption; and second, *for* imperatives of labor and human rights. Thus, this chapter will discuss the pervasiveness of mobile hardware and software, the development of mobile applications, and our persistent encouragement to upgrade, in light of *both* the enclosure and privatization of education, *and* the commodification of content (Jarvis, 2010). This will be related to global social justice and ethical imperatives, including the labor rights, resource accumulation, geographical dispossession, and supply chains in which the uses of mobile technologies, especially in the global North, are implicated.

Each of these interconnected ideas implicates and enmeshes the use of mobile technologies within the webs of a transnational, capitalist market that has been described as "Empire" (Deleuze & Guattari, 1984; Hardt & Negri, 2000). Yet these webs of capitalism and transnational power relationships keep those who notionally benefit from access to mobile technologies and mobile learning at a distance from the effects of their consumption. This chapter therefore opens up a discussion of what might be done in opposition to the hegemonic logic of capital as revealed in Empire. This will focus upon developing a public, sociotechnical critique within education that is based upon an ethical, digital literacy, which in turn frames open-sourced, community designed and implemented technologies as solutions to problems that are defined cooperatively.

The Perceived Value of Mobile Learning

The limits of analyses of mobile learning in tertiary and higher education are illustrated ably by the Joint Information Systems Committee (JISC) (2012) "mobile learning infokit." Produced by the public body charged with overseeing technology use in UK tertiary and higher education, this set of publications indicates how approaches to mobile technologies are legitimized at a national level by organizations that support and influence universities. These publications are interesting in that they take the outcomes from stakeholder analyses and horizon scanning, in order to develop a strategy and set of diagnostics for implementing institutional responses to mobile learning.

The focus is on student engagement and for responsiveness to the needs of students for flexible access to resources and communication on their own devices. Thus, JISC (2012, n.p.) states that

> without an overall vision for what mobile learning is able to achieve in your particular context it is doomed to failure. [Institutions need to] consider what learners are increasingly coming to expect in terms of their ability to use mobile devices for anytime, anywhere learning. It is not enough to simply engage and cater for learners, however, as staff must also be on board with initiatives. Cultural considerations are important for any change management project and particularly when technology is involved that is more often used for social reasons. Finally, whilst the novelty factor may enable an organization to gain some initial traction, it is important that mobile learning initiatives are sustainable.

Sustainability in this context is related to localized cultural and institutional considerations that are sympathetic to customs and usage, pedagogies and scholarly expectations. It is an agenda that has been related to discussions of curriculum personalization and to the development of digital literacies by students and staff, alongside a desire to enable students to develop their life and employability skills for knowledge work. Thus, the perception is that mobility, and the flexibility that it entails, is central to the ability of individuals to connect their identities as scholars or agents to the work of institutions, like universities, colleges, or employers, through personal technologies. Such flexibility underpins sustainable curricula and is further related to issues of scale and scope, whether in the global North in terms of work-based or placement learning or the global South in terms of access to what are regarded as basic amenities and information (Gelb & Decker, 2012).

As a result, the flexibility and efficiency of access to both information and communication, whether via the open web or through application-driven content or inside fluid social networks, is seen to be a core characteristic of mobile learning. However, this is also an agenda that is being driven by value as it is described economically. Thus, in extending its statement on sustainability JISC (2012, n.p.) notes,

> "Sustainability" means different things to different groups, to finance and marketing teams the focus is upon Cost/benefit issues; to IT personnel it is about keeping systems up-to-date; and to academic staff and students it is about the technical systems remaining relevant to desired pedagogical (and social) outcomes.

This economic agenda for mobile learning grows from the wider focus within UK tertiary and higher education on technology for efficiency gains, and the desire to see increased mechanization as a means of extracting value, or for reducing the circulation costs of capital. Thus, the Higher Education Funding Council for England (HEFCE, 2012) focuses on technological innovations for cost reductions, business-process reengineering, and efficiency gains, which themselves might underpin the radical transformation of the university as a "business" (HM Treasury, 2012), in light of structural changes across education that are political, financial, technological, and competitive. Mobile learning is an increasingly important node in this restructuring of education around personalization and efficiency.

One outcome is that the discourse of sustainability that is legitimized by publicly funded bodies connects technological innovations to the policy drive for economic impact across society, alongside various commercial imperatives. Thus, the rationale of the UNESCO (2007) "Information and Communication Technology (ICT) toolkit" for educational policymakers, planners, and practitioners demonstrates the complexity of the issues that emerge from any technological implementation and that amplify the tensions that are catalyzed by the use of personal, mobile technologies in social and commercial spaces:

> Educational authorities are under tremendous pressure to provide every classroom (if not every student) with technologies, including computers and their accessories and connectivity to the internet. The pressures are coming from vendors who wish to sell the most advanced technologies, from parents who want to ensure that their children are not left behind in the technological revolution, businesses who want to replicate in schools the dramatic impact that ICTs have had in the worlds of commerce, business and entertainment, and from technology advocates who see ICTs as the latest hope to reform education. (UNESCO, 2007, n.p.)

This then highlights the increasingly complex interplay between personal, educational and work contexts, identities and technologies, and the tensions that may arise in terms of the use and ownership of data, security, privacy, and identity management by public or private bodies (Bush, 2012). As a result the connections between educational innovations, commercial imperatives, and fiscal realities are placed symmetrically alongside the desire to enhance student learning and to transform institutions. These aligned sets of outcomes are generally revealed as potentially transformatory in terms of mobile learning because they connect business efficiency directly to personalized access; to flexibility of delivery; to opportunities for work-based and placement learning; and for engaging marginalized students.

Thus, considerable evidence is being mounted for the potential of mobile learning to underpin business transformation, to engage public and commercial sectors in partnership, and to effect change in pedagogic practices, with the dominant discourse being refocused upon mobile technologies as one of the latest hopes to reform or transform education. However, there has been limited critical engagement with the production and consumption of *either* those pedagogic approaches *or* the technologies themselves, in terms of the reality of labor in capitalism. It is from inside this space—from a perspective that is less about the perceived value of mobile learning and more about how mobile learning enables the reproduction of alienating capitalist social relationships—that a meaningful critique might be developed, and a process of refusal enacted.

Mobile Learning and Labor in Capitalism

A materialist analysis of the relationships between machines and humanity highlights how capital uses technologies to reshape and redefine the relationships between humanity, nature or the world, and power (Marx, 2004). Marx (1993, p. 594) argued that technologies in the form of machines "are the materialized power of knowledge." This materialized power reflects the relationships that exist between those who use technologies to create, repurpose, and reproduce society, and *both* those who innovate around those specific technologies (like the providers of applications) *and* those who use them in their labor (like educators). This also reveals technologies as sites of social struggle through which hegemonic positions are developed, legitimated, reproduced, and challenged. This revelation was important for Feenberg (1999, p. 87) who subsequently argued for "a critical theory of technology [that] can uncover that horizon, demystify the illusion of technical necessity, and expose the relativity of the prevailing technical choices."

For Mackinnon (2011) this struggle is important because "a substantial, if not critical amount of our political discourse has moved into the digital realm. This realm is largely made up of virtual spaces that are created, owned and operated by the private sector." The control of tools and spaces for deliberation, sharing, and reproduction, where controversies can be played out, is compromised by the role of those with "power-over" labor (Holloway, 2002). This power-over is not only realized in the form of wage labor, but is also revealed through the control of access to information and communication or in its legitimation. In Nolin's (2010) terms, this opens up as a threat to the distributed, social web through the ideologies of speedism, boxism, and markism, whereby corporations seek to control and regulate *both* the layers on which the internet is built and distributed, *and* the appliances on which web-based artifacts and products are accessed.

This marketized, ideological threat of the enclosure of the open web through its commodification amplifies Feenberg's (1999) position that technology is a critical site of sociopolitical struggle, beyond relatively neutral issues like participation, access, and engagement. Thus, as humanity is entwined and embedded with personalized, technological appendages, the possibilities for cybernetic control and the further individuated alienation of subjectivity become more apparent (Habermas, 1987; Tiqqun, 2001). This reflects the amplified alienation of labor inside the "social factory," achieved through the symbiosis of human and machinery, and the capture and marketization of the entirety of the lifeworld for value extraction (Negri, 1989). For Harvey (1990, p. 157), such alienation is an outcome of the response of neoliberalism to the economic and political crises of the 1970s, through which capital actively sought new strategies that "put a premium on 'smart' and innovative entrepreneurialism." Hardt and Negri (2000, p. 406) then see the development of smart technologies as a deeply political antagonism, for "machines and technologies are not neutral and independent entities. They are bio political tools deployed in specific regimes of production, which facilitate certain practices and prohibit others." In this critique, mobile technologies are critical as they enable forms of human-machine symbiosis that in turn enable capital to fuse or augment humanity with fixed capital, in order to connect dead and living labor, and thereby *both* develop new ways of extracting value *and* reduce the circulation costs of capital (Marx, 2006).

Innovation around mobile, identity-driven technologies further objectifies social relationships as commodities from which value can be extracted. This can take place, for instance, through the monitoring and harvesting of personal and geospatially tagged data, the enclosure and control of spaces or applications of consumption, the use of venture capitalism to support specific mobile networks and technologies, and the use of mobile technologies for the technological augmentation and capture of affectivity. As a result, *both* the anytime/anywhere capabilities of mobile technologies, *and* their identity-driven, personalizable reality, enable the real subsumption of everyday activity inside the reproduction of capital. This ensures that for an individual,

> the creative power of his labor establishes itself as the power of capital, as an alien power confronting him...Thus all the progress of civilization, or in other words every increase in the powers of social production...in the productive powers of labor itself—such as results from science,

inventions, divisions and combinations of labor, improved means of communication, creation of the world market, machinery etc., enriches not the worker, but rather capital; hence only magnifies again the power dominating over labor...the objective power standing over labor. (Marx, 1993, pp. 307–308)

Mobile technologies, as a site of affective and augmented value, therefore could be said to act as a transformatory force but only in terms of supporting capital in its destruction of "all the barriers which hem in the development of the forces of production, the expansion of needs, the all-sided development of production, and the exploitation and exchange of natural and mental forces" (Marx, 1993, p. 409). This exploitation is constantly seeking to overcome the barriers that result from physical limitations, and increasingly rests on the fusion of the human as social being with technology or technological applications, in order to create new commodities that overcome the limits of underconsumption (Žižek, 2009).

The push for engagement with mobile technologies is especially important in the context of capital's need to innovate in order to create high marginal productivity from which to drive further commodification and overcome barriers to consumption (Clarke, 1994; Marx, 2004). Personalization amplifies this process through the customization and the generation of individuated desires that are fetishized in the machine or application or network. However, over time "moral depreciation" affects the gains made by technological innovation. Thus,

in addition to the material wear and tear, a machine also undergoes, what we may call a moral depreciation. It loses exchange-value, either by machines of the same sort being produced cheaper than it, or by better machines entering into competition with it. In both cases, be the machine ever so young and full of life, its value is no longer determined by the labor actually materialized in it, but by the labor-time requisite to reproduce either it or the better machine. It has, therefore, lost value more or less. The shorter the period taken to reproduce its total value, the less is the danger of moral depreciation; and the longer the working-day, the shorter is that period. When machinery is first introduced into an industry, new methods of reproducing it more cheaply follow blow upon blow, and so do improvements, that not only affect individual parts and details of the machine, but its entire build. It is, therefore, in the early days of the life of machinery that this special incentive to the prolongation of the working-day makes itself felt most acutely. (Marx, 2004, p. 528)

Under the logic of competition and in order to maintain increases in the rate of profit, the impulse is for constant innovation across a whole socio-technical system. This drive focuses upon maintaining or increasing the rate of extraction of relative surplus value, in part by removing the barriers of underconsumption. Mobile technologies are important in this scenario because they enable a number of shifts—such as the working day to be extended, work to be embedded throughout the whole of social life, the capture and accumulation of affects and personalized data, and a reduction of the costs of the reproduction of the value of the machine. The reduced costs of production of these technologies, through assembly that is generally conducted in the low-wage economies of the global South, and the high value of marginal innovations related to the production and distribution of applications, then amplifies the persistent drive to renew mobile technologies, and the concomitant needs, to produce, market and regulate new applications or the mobile web; to migrate services to the cloud; to control and mine data; and to manage identities.

For capital, capturing and mining this type of activity is an important field of innovation and value extraction, not least because "data suggest that less than 10 percent of human life is completely unemotional. The rest involves emotion of some sort" (Cowie, 2005, p. 1). Thus, capturing emotionality or affect through personalized technologies focuses upon enhancing "the quality of human-computer communication and improving the intelligence of the computer" (Tao & Tan, 2005, p. 981). Thus, mobile hardware enables virtual and augmented reality applications to become more immersive and integrated into daily life. For instance, smartphones and their applications extend the marketization of everyday experience through the enclosure of content and concomitant subscription or rental charges. These technologies enable the further commodification of a lifeworld through augmented objects or as augmented information is projected onto real-life contexts (Zhou, Been-Lirn & Billinghurst, 2008).

From this process, two elements emerge as central in understanding mobile learning as it relates to labor in capitalism. First, capital utilizes these technologies to enclose and commodify an increasingly fluid and identity-driven set of social relations, which can be controlled and mined, and then form the basis of further exchange (Virno, 2004), catalyzed by work inside the university and based on mutations of human subjectivity. Second, capital commodifies and extracts value from everyday experiences and relationships in personalized networks that are "always on," in order to reduce the unproductive circulation time of capital, and thereby increase the rate of profit and relative surplus value. In this process of immaterial reproduction, social relations are increasingly structured by technocratic, sociopolitical hierarchies that are increasingly coercive and exploitative, using technologies that are inserted into the everyday activities and lifeworlds of living human subjects. This necessity reveals the real subsumption of labor under capitalism in its performance of real-world tasks. By supplementing an everyday reality with personalized, mobile, and immersive virtual objects or data, the immaterial laborer is able to produce and consume new services that move capitalist reproduction beyond the barriers of underconsumption. As a result, the emergent technologies cultivated through innovation and imposed as structure by neoliberalism form a network designed for the subordination of all activity to the law of value (Harvey, 2010). This is a relentless dynamic, centered on capitalism's constant revolutionizing of the means of production, in order that capital can drive "beyond every spatial barrier... [and the ability to enhance] the creation of the physical conditions of exchange—of the means of communication and transport—the annihilation of space by time—becomes an extraordinary necessity for it" (Marx, 1993, pp. 524–525). As Marx highlights (1993), the development of such technologies that subsume all of human life under capital's logic strengthens the idea that capitalist relations are natural and purely technical. This positivism and the sociotechnical systems that infuse it then legitimize the analyses, practices, and conceptions that wrap around mobile technologies and mobile learning.

Defining a Politics of Mobile Learning: Against Empire

Innovation inside education is framed by the politics of power that are revealed inside capitalism. For instance, claims are made for educational innovation tied to sustainability, or equality of opportunity, or social inclusion, without a critique of how the history of labor in capitalism informs the development of these values. Thus, educators are unable to take a systemic view and are unable to separate out humanist values from the need to create surplus value as the primary form of social mediation within capitalism. Educational values are predicated instrumentally on the tenets of liberal democracy, on tropes of equality or liberty, or on often ill-defined practices/qualities like respect or openness. Thus, even inside the university it becomes impossible to think otherwise, and to imagine a different form of social life beyond the realities of capitalist work.

In attempting a more meaningful critique of educational innovation with mobile technologies, it is important to locate education inside the emerging discussion of "polyarchy" (Davies, 2012). Polyarchy is an attempt to define an elitist form of democracy that would be manageable in a modern society.

It focuses upon normalizing what can be fought for politically, in terms of organizational contestation through free and fair elections, the right to participate and contest offices, and the right to freedom of speech and to form organizations. This forms a set of universal, transhistorical norms, through which it is simply not acceptable to argue for other forms of value or organization beyond democratic capitalism, without appearing to be a dissident or agitator. Within the structures of polyarchy, it no longer becomes possible to address the structural dominance of elites within capitalism, or the limited, procedural definition of democracy inside capitalism. Compounding this political enclosure is the control of the parameters of discussions about values or value relationships like democracy and equality, or power and class.

Important here then is to understand how innovation flowing to/from the university supports the ways in which neoliberal capitalism intentionally designs, promotes, and manages forms of democracy and governance that complement its material objectives, limit participation and power sharing, and support coercion (Harvey, 2010). Thus, the rhetoric of student entitlement and student-as-consumer enables the market to penetrate the education sector, in order to open its resources up to the dominant or hegemonic order, and to manufacture consent for commodified practices (Ball, 2012). Manufacturing this consent depends upon the coercion of organizational leaders, and the extension of the argument that there is no socioeconomic or political alternative. For the extension of polyarchy, it is critical that once economic and productive power has been extended into, for instance, the educational space, that domination extends to the political-, social-, and class-based relations in that space. This is achieved through the implementation of ideological control inside the sociotechnical institutions and cultures of civil society, which in turn make it impossible to step beyond the controlling logic of the rights of consumers.

It is from inside this bounded, polyarchic space that a critique of mobile learning might be situated, in order to identify opportunities for dissent, negation, and pushing back against the alienating rhetoric of capitalist work (Holloway, 2010). This critique emerges from two strands, first, in being *against* pedagogies of consumption that define mobile learning through the commodification of engagement and activity; second, from the recognition that mobile technologies, and hence mobile learning, help to recover global imperatives of labor and human rights.

Against a Pedagogy of Consumption

Educators are not immunized from this process of commodification, and scholarly works (in terms of courses, technologies, knowledges, and cultural assets) now form sites for the extraction of surplus value inside an education market (Ball, 2012). Mobile learning and mobile technologies are one space in which competition between teachers and institutions can be structured socially, and this competition informs the allocation/abundance of relevant academic labor. Such labor or knowledge work inside both tertiary and higher education is particularly valuable as a result of the amount of socially necessary labor time embedded in its academic products. Inside education, the specialization of the work and the skill levels required to innovate promise high rates of surplus value extraction, especially where technological research and development catalyzes efficiencies in production and a reduced circulation time for specific commodities (Clarke, 1994).

This specialization and the promise of increased productivity fuels the drive for mobility and flexibility in agendas that frame the skills developed in education by the needs of the labor market. In the face of persistent technological renewal, digital skills become collateralized as use-values for specific individuals. These use-values are then commodified inside institutions as exchange values that are valorized through so-called anytime/anywhere access. This is more important as education becomes increasingly personalized, marketized, and commodified, and as spin-off intellectual capital or knowledge work increases in value. Thus, the nature of exchange inside and across the social factory enables capital to develop a marketized and co-opted academic process. Inside the tertiary and higher education sectors, the struggle between labor and capital lies in the creation and commodification of cognitive capital, and in the production and circulation of "immateriality," where individual emotions and affects, cultural cues and mores, and the construction of the relations between individuals "are themselves the very material of our everyday exploitation" (Žižek, 2009, p. 139).

The pervasiveness of mobile hardware and software, and the persistent desire to upgrade that is catalyzed by branding and marketing, risks further privatizing the contexts of education. Thus, educators might usefully ask whether a focus on mobile applications, as opposed to the development of the mobile web, reinforces a pedagogy of consumption through the commodification of content that is defined in large part by corporations or publishers, and which underpins the extraction of value through rent. This is based, in part, on transnational software and hardware corporations driving content-based innovations that enclose and threaten the idea of the open web, within the context of their brand and procurement processes, and the dominance of their cultural perspectives. Moreover, in promoting engagement with an application-driven market that is either enclosed or for profit, educators risk surrendering both pedagogy and curriculum delivery to the impulse to search for the latest software innovation or hardware upgrade.

This fetishization of the next innovative tablet, smartphone, or application, risks surrendering the power over our access to and production of learning resources to corporations. Moreover, it frames a world where the consumption of prepackaged or prevalidated content is an idealized norm, as opposed to the ability for students or academics to utilize machines to produce, enhance, challenge, or reform their social relationships.

As a direct result of their user-generated characteristics and the ability to monitor activity, capital is increasingly able to utilize mobile tools to facilitate remote, placement, or workplace learning, and proletarianized labor at a distance from any formal, Taylorized setting. This, in turn, enables capital to distribute available commodity and leveraged skills among low-wage societies through outsourcing, and the evolution of cloud computing, application service provision, and Software as a Service. For De Saulles and Horner (2011), this model of engagement with mobile technologies is made more ethically problematic because those tools enable decentralized surveillance of activity by the state, by network carriers and/or corporations inserted inside educational activities, by institutions like schools, or by individuals operating inside dispersed social networks. Thus, the increasingly converged and user-generated characteristics of untethered machines, which focus upon "high-tech" capabilities such as WiFi/4G networks, video/audio capture, GPS capabilities, and so on, impact socioethical imperatives, as they relate to education through, for instance, monitoring and performance management, privacy concerns and data mining.

These capabilities for the surveillance of remote learning, amplified through mobile technology, enables the dynamics of cognitive labor to pervade higher education inside the social factory and thereby amplify immateriality on a global scale. In particular, the deep interconnections between data mining, mobile technologies, and cloud computing enables corporations to sell services into education that enable it to extract value from social networks and personal interconnections through the corporate control of systems, networks, and data, *and* to reduce the circulation costs of productive capital through scalable and elastic digitally enabled capabilities that are delivered as a service from low-wage circuits into those spaces from where high value can be extracted.

As Greef, Kees van Dongen, Grootjen, and Lindenberg (2007, p. 1) argue in relation to augmented cognition, the aim for engaging with mobile learning inside capitalism is "the creation of adaptive human-machine collaboration that continually optimizes performance of the human-machine system." Using mobile, untethered tools the student is optimized as labor power and her very humanity is alienated from her being, as commodity or appendage in a cybernetic system (Tiqqun, 2001), which in turn appears inherent or natural to capital. As a result, mobile learning connects an emerging human-machine symbiosis through the structures of education, to the production and consumption of a lifeworld inside capitalism that is increasingly dependent and reliant on personal technologies, inside a system that legitimizes an increasingly controlled set of activities. The very fact of capital's increasing enclosure of the human body and human affects inside its machinery of exploitation means that education's very lifeworld is a site of surplus value creation, extraction, and accumulation.

For Labor and Human Rights

The labor rights, resource accumulation, geographical dispossession, and supply chains that underpin the means of production and distribution of mobile technologies are missing from a critique of mobile learning. An analysis might be tied into the human and labor rights of those engaged *both* in mining the resources that enable mobile telephony to scale efficiently *and* in the assembly of those products. These abuses are connected through webs of transnational global finance, mining corporations and media firms to the educational practices that are increasingly common in the global North, and which form revelations of the imperatives of capital.

The historical and material genesis and impact of the war in the Democratic Republic of Congo (DRC) has been related to the trade in rare earth minerals (Eichstaedt, 2011). In over a decade of fighting, more than five million people have been killed, and Amnesty International have documented human rights' abuses including gang-rape, mutilation, enforced disappearances, and the militarization of boys and young adults. In this war, Eichstaedt (2011) emphasizes the material importance of the DRC's mineral deposits, and in particular coltan, which is verbal shorthand for columbium and tantalum. Coltan is critical in electronics because it is very difficult to corrode, effectively resists heat, has superconductivity, and can handle huge electric charges. Moreover, coltan is generally found in close proximity to tin, which forms a primary source of solder in electronic devices.

In a financially poor area of the world, Eichstaedt (2011, p. 111) argues that "demand for these minerals stoked the flames of war as exploitation grew." This point develops the argument of Ware (2001, n.p.) that

> coltan is increasingly exploited in the mountains in the conflict torn eastern part of the country. The Rwanda and Uganda backed rebels have primary control over the ore and are reaping huge profits which maintain and finance the protracted war. It is estimated that the Rwandan army made $20 million per month mining coltan in 2000. As coltan is

necessary for the high-tech industry and as demand increases, motivation to pull out of the DRC by Rwanda, Uganda, and Burundi decreases.

Thus, despite the relatively small role that tin and coltan from the DRC play in the global market for rare earth metals, the revenues flowing from the control of mines in the east of the country is hugely significant in terms of local geopolitics. Eichstaedt (2011, p. 143) notes "that significance can be counted in the millions of dollars and the millions of lives lost or damaged over the past sixty-five years in the worst human death toll since World War II."

Mining these minerals, which flow into the global market for the means of production of mobile technologies, highlights the complex, transnational interrelationships and networks between armed militias, gold and mineral trading networks, mineral exporting companies, global smelting corporations, and telecommunications companies. In part through mobile technologies, educational institutions and networks in the global North are connected to these transnational networks. These interconnected networks of power were highlighted in the UN Security Council Experts' Report (Mahatani, Debelle, Diallo, Dietrich & Gramizzi, 2009) that linked the demand from the emerging global market for electronics to the human rights' abuses witnessed in the DRC, and to extensive networks of arms dealers, military officials, and diplomats (Eichstaedt, 2011). So, it is argued that demand for these minerals from the global North are the driving forces for war, and that those who benefit are multinational corporations involved in Western high-tech innovation and development. For Eichstaedt (2011, p. 113), this resulted in "a profusion of militias wreaking havoc wherever they went, killing any who might counter their control of the mines." Indeed, Eichstaedt (2011, p. 54) quotes an official who states, "The Congo has resources that attract [buyers] who can go with the government or with armed groups...The rebels are exploiting minerals. They're sending children into the holes to dig [minerals]. Yet no taxes are left behind."

The complexities of local geopolitics, of shifting national boundaries and ethnic tensions exacerbated by access to natural resources and mineral wealth, and of the nature of a globalized market for those resources inside the commodities that fuel technological innovation in the global North, mean that making sense of appropriate action is highly problematic. The matrices of interrelationships that build up around *comptoirs* (trading houses) and *négociants* (middlemen), traders, smelters, and global mining companies, through to national governments and technology corporations means that supply-chain management and assurance is incredibly laborious. However, Global Witness (2009, p. 4) argued,

In their broader struggle to seize economic political and military power, all the main warring parties have carried out the most horrific human rights abuses, including widespread killings of unarmed civilians, rape, torture and looting, recruitment of child soldiers to fight in their ranks, and forced displacement of hundreds of thousands of people. The lure of eastern Congo's mineral riches is one of the factors spurring them on. By the time these minerals reach their ultimate destinations—the international markets in Europe, Asia, North America and elsewhere—their origin, and the suffering caused by this trade, has long been forgotten.

Eichstaedt (2011) raises issues of certification and designation of minerals and mines, and the processes of due diligence that led to bills against the trade in conflict minerals being ratified by the US Congress. The UN Security Council Experts' Report (Mahatani et al., 2009) similarly argued for enhanced due diligence processes to be carried out cooperatively and in public. Yet in terms of the global supply of rare earth metals, it is important to note that only a small amount comes from the DRC, which means that for transnational corporations, invoking due diligence policies for these mines is not worth the cost. Thus, there is little incentive for those corporations to invest in tracking systems or in maintaining the mines, and their withdrawal means that miners will be left without incomes or placed at the mercy of militias and less scrupulous governments. At issue then is the extent to which educators who are framing a demand for mobile learning are implicated in the war in the DRC, through their relationships as consumers or promoters of the hardware of multinational companies that may source conflict minerals. For Eichstaedt (2011, pp. 214–216), it is here that the personal becomes political and might underpin action.

> We all use and depend on all sorts of high-tech devices in our daily lives…We are all linked on our shrinking planet…Forming personal and lasting bonds with people is the most effective and powerful way to effect change…Feet on the ground, followed by time, toughness, and commitment to change is needed. Nothing less.

This approach also connects the consumers of mobile technologies to the manufacture of those tools. In particular, there has been a focus upon working practices and the conditions of labor in factories in which FoxConn has produced the iPad, iPod, and iPhone in China for Apple. In analyzing labor rights, the use of mobile technologies in the global North has been implicated inside transnational, high-tech networks, which it is claimed feed alienating behaviors. Thus, there are reports of game farming in virtual sweatshops in the Far East for Western clients, and of alleged tax avoidance by mobile phone operators.

Yet these webs of capital, with its transnational circuits of raw materials, value, and power, keep those who notionally benefit from the immiseration of others at a distance from the effects of their consumption and the effects of demand on supply-chain mechanics are not discussed. Instead, the spectacle becomes focused upon marketing imperatives based around corporate hegemonies or business models, transnational power or nationalism. More occasionally it is about how our Western, liberal data rights are being infringed, or about how the police are using mobile technology to target protesters, or upon the impact of mobile technologies on bee populations. There is little or no engagement with the structures that connect global consumers to the supply chains for and production of these technologies.

Against Empire

Virno (2001) argues that capital keeps the structural realities of the world at a safe distance, so that excessive consumption, equivocation concerning morality and the ridicule of marginalized voices can enable endless, repetitive practices of commodification in a world where there are seen to be no alternatives. In this polyarchic space, cynicism becomes the defining feature of the emotional situation of global politics. The argument becomes, what on earth can we do about such powerlessness and distress, when our power is apparently limited? One of the issues here is the functioning of what Hardt and Negri (2000) have called Empire, the metamorphosis of capital into a new planetary regime in which its economic, military, administrative, and communicative components combine into a system of power "with no outside" (Meiksins-Wood, 1997). Thus, Empire is a twenty-first-century critique of global capital, which now taps its subjects as labor power and also as consumers, learners, and raw materials. Deleuze and Guattari (1984) have also argued the case that capitalism is now a planetary "production machine," assembled from flows of labor, finance, and technology, where the quest for profit drives new technical machines, new products and practices, cracks old habits, and throws all bounded domains or territories (that are geographic, social, and subjective) into upheaval. It then reterritorializes these domains through enclosure, policing, and commodification.

For these critics, Empire forms a regime of biopower, capable of exploiting social, subjective, and biological life in its entirety, and for profit (Hardt & Negri, 2000; Holloway, 2010). So Empire and the transnational corporations that form nodes of power within it, and whose networks are circuits for accumulation and profit, covers all of human life, though marketing, gameplay, work, privatization of public assets, data mining, advertising, the constant renewal and upgrades of mobile technologies, and so on. It is these networks that then underpin "immaterial labor," through the commodification of our desire for play or for the latest cheap, powerful, miniaturized device. Thus, for instance, the much-heralded "Raspberry Pi" is connected to the desire to engage young people in programming through affordable, flexible, mobile devices that reveal the inner workings of the machine as it relates to programming. Yet, there has been little discussion of the component parts that make up the machinery, and how they are sourced. This machine uses a "Broadcom" corporation bcm2835 SoC (system-on-a-chip). According to a company engagement report made by the Triodos ethical bank, Broadcom was ineligible for ethical investment during that financial year because of their performance regarding conflict minerals, cooperation with repressive regimes, and human rights (Triodos, 2011).

Thus, these interconnected webs of mining and trading corporations, manufacturing, transnational finance, and high-technology industries, reveal what might be termed an "Empire of things." In this Empire, social life is geared around the interconnected production and consumption of material life as mediated through increasingly personalized technologies. This set of transnational processes is supported by a socially diffuse intellectuality and set of desires, which is in turn generated by a vast educational apparatus that wishes to connect, for example, mobility, pedagogy, play, and labor, through mobile or personalized technologies. As a result, through a focus on practices of consumption and an ignorance of human or labor rights, educators in the global North are implicated and enmeshed within the alienating webs of a global, imperial market. However, Dyer Witheford and de Peuter (2009) argue that while devices are enslaving, this is not to deny that they are pleasurable. Thus, educators and users need to recognize how that pleasure itself channels power, and how they might critique the realities of their uses of technology, in order to imagine alternative narratives and practices.

What Is to Be Done? For an Ethical Digital Literacy

The Ethical Issues of Emerging ICT Applications (ETICA) project scoped the interplay between sociotechnical systems and ethical applications. The project team argued that a technology

is a high level system that affects the way humans interact with the world. This means that one technology in most cases can comprise numerous artifacts and be applied in many different situations. It needs to be associated with a vision that embodies specific views of humans and their role in the world. (Ikonen, Kanerva, Kouri, Stahl & Kutoma, 2010, pp. 3–4)

This connection between technology and both humanity and morality is amplified by Harvey's (2010) engagement with Marx's (2004, p. 493) fragment on machines, in which the latter described technology's place inside a historical totality:

> Technology discloses man's mode of dealing with Nature, the process of production by which he sustains his life, and thereby also lays bare the mode of formation of his social relations, and of the mental conceptions that flow from them.

In this view, mobile technologies enable the reproduction of "man's modes" of recasting and reforming social relationships. Through an engagement with this fragment, Harvey (2010) suggests that there are seven activity areas that underpin meaningful social change, including issues of conducts, practices, and relationships between people acting in the world—all of which might include ethical forms. These are as follows:

1. Technological and organizational forms of production, exchange, and consumption
2. Relations to nature and the environment
3. Social relations between people
4. Mental conceptions of the world, embracing knowledges, cultural understandings, and beliefs
5. Labor processes and production of specific goods, geographies, services, or affects
6. Institutional, legal, and governmental arrangements
7. The conduct of daily life that underpins social reproduction

Harvey (2010) highlights how technology is tied to organizational forms, both in and beyond the institution, and that it is impacted by social, environmental, and political-economic effects, relations, and conceptions. This places a critique of mobile technology firmly within a broader range of life experiences in the lived, capitalist world. Moreover, the nature of these lived experiences is social, and not personal, in spite of our hopes for engagement with the personalized aspect of mobile learning and technologies. Achieving holistic social change is a complex task, and beyond the instrumental potential of an engagement with technology alone.

Thus, as Eichstaedt (2011) notes, global solutions are required if educators are to attempt to move beyond the problematized, polyarchic implementation of mobile technologies that is described above. In doing so, it is important that educators are able to address the structural, rather than instrumental, issues that arise from their pedagogic practices with mobile learning. That said, it could be argued that educators are well placed to discuss the following questions within their institutions, networks, and curricula, in order to consider meaningful ethical change:

• How do educators, policymakers, and educational technologists lobby mobile and high-technology corporations, vendors, resellers, and commissioners, in order that they justify the extraction of raw materials, and the production processes that they use as means of production for mobile technologies? How might this be achieved through cooperation, mutuality or in association with others and throughout their daily work?

• How do educators, policymakers, and educational technologists work for technological decisions, like procurement and the development of applications and hardware, to be based on community need related to a critical analysis of socioenvironmental impact and human rights, rather than on a discourse of cost-effectiveness, consumption, monetization, economic value, and efficiency?

• How do educators, policymakers, and educational technologists lobby for consensus in engaging with open systems architectures, which might be focused upon open-sourced, community designed and implemented mobile technologies? Can these personalized, mobile technologies be mapped to a deliberative, associational pedagogy, whereby the personal and affective might be judged alongside competing priorities, for instance labor and human rights?

• How do educators, policymakers, and educational technologists work for a digital or technological literacy that is ethical across full range of their institutions, networks, and curricula? How do those with a clear stake in education work up an ethics of mobile learning that is against alienating human and labor rights and for meaningful social change?

Part of the response to these questions might be shaped by a critique of the transnational politics and power of Empire that emerges from inside counterhierarchies (Davies, 2012), as they are revealed by the politics of mobile learning. Flowing from this interplay, different power and political configurations become possible, and as a result new hegemonies become possible as cracks in political and civil society (Holloway, 2010). Developing these new counterhegemonies or alternative spaces *both* for organizing civil society *and* for imaging new forms of value, depends not upon the market or the rights of consumers, but on human consciousness and human relationships. The inside of educational spaces might offer one context for defining such new, socialized forms of value.

Thus, a structural engagement with mobile learning, through the realities of their location inside capitalist work, offers the possibility that educators might develop humanizing, counterhegemonic political spaces, through the creation of participative, democratic pedagogies. Such pedagogies enable critiques of power and political economy, where they are explicitly revealed in the production and consumption of mobile technologies in everyday pedagogic practices. This work might involve the following types of activities:

1. A refusal to engage in dehumanizing pedagogies of consumption, and as a result pushing back against the reproduction of the alienating social relationships of labor in capitalism, through an active engagement with open-source or open web applications. One result might be that educators and students are actively encouraged to produce and share open curricula and artifacts in ways that reveal humanizing engagements with transnational issues of human and labor rights, and our collective access to the means of production. These mobile-enabled artifacts and pedagogic practices solutions do not form new commodities, but help maintain the diversity of expertise in a community and help connect modular networks together as a means of deconstructing power.

2. The increased prioritization of mobile technologies in institutionalized and networked strategies for associational democracy, modular community building, and the diversity of skills sharing and development. Mobile learning and the pedagogies that underpin their use thereby focus upon understanding mutuality.

3. Educational networks act as mobile hubs for community-level engagement with technologies, and for accessing high-level digital processes. Networks of mobile technologies are used for maintaining the diversity of skill sets within and across communities, rather than simply enabling modes of consumption.

4. The relevance of marginal developments like application-based, locative, and augmented reality services is questioned through consensus, and related to social need and issues of privacy and identity.

5. Persistent and ongoing procurement and renewal of mobile hardware and software is rejected, in favor of reuse and repurposing.

6. The use of mobile technologies and their connection into webs of open education rejects a postcolonial discourse focused upon new markets and new consumers, for associational democracy, the sharing of communal expertise, and the coproduction of everyday practices and lifeworlds.

Discussing how these types of practices are enabled through mobile technologies might also enable new forms of resistance that are against polyarchy to be developed. The question is how to turn the possibilities enabled through a critique of the manufacture of mobile technologies and their place in mobile learning into meaningful counterhegemonic practices that can resist, push back against, and overturn the imperatives of Empire. It may be that the starting point is an ethical digital literacy that critiques the place of mobile technologies inside an educational space that is itself defined socially. In his novel *Nostromo*, Joseph Conrad (1963) wrote about the social and material history of the Congolese, as their land was despoiled and as they were colonized in the nineteenth century:

> There is no peace and no rest in the development of material interests. They have their law, and their justice. But it is founded on expediency, and is inhuman; it is without rectitude, without the continuity and the force that can be found only in a moral principle.

Our current use of mobile technologies needs to be recast in light of a critical history of their production and consumption, in order to imagine alternatives that reclaim humanity from the fetishization of an Empire of things.

CHAPTER 10

Tweak: Biosocial Imaginations and Educational Futures

Nick Lee and Johanna Motzkau

Introduction

Imaginative and practical links have long been drawn between children's life processes and projected futures (Rose, 1989; Turmel, 2008). Many educational enterprises have attempted to take hold of the complexity of human maturation so as to bend it to chosen purposes, such as the promotion of literacy or increasing independence in learning. Thus, education can be seen as a set of practical "biosocial" activities (Lee & Motzkau, 2012) aimed at forging functional connections between life processes and socially preferred abilities and qualities. These activities have a defensible tendency to anticipate futures and to prefigure, and so limit and enable, children's future behaviors, capabilities, and values. Such activities have both been guided by and have given rise to metaphors and narrative structures that try to frame and make sense of the complexity of the human condition. Consider the power of a phrase like "as the twig is bent, so the tree shall grow" in figuring children's mental, physical, and moral development as targets of intervention, or the imagined adult careers that, as part of some children's projected autobiographies, can motivate committed study. In the terms of this chapter, such metaphorical and narrative frames comprise "biosocial imaginations" (Lee & Motzkau, 2012). The key question addressed, therefore, is the increasing use of biotechnologies to achieve these preferred biosocial outcomes.

In what follows, we will focus on just one of a range of biosocial imaginations that feature in current attempts to shape and to influence educational futures. For the sake of brevity, we call it "tweaking." In our terms, to "tweak" is to make a subtle adjustment in the hope of dramatically improving outcomes. The term has resonances with two other contemporary ways of imagining the tractability of human behavior and capability; the "nudge" (Thaler & Sunstein, 2009) that has emerged from the field of behavioral economics to offer governments' strategies for behavior change in populations while circumventing the charge of paternalism; and, the "hack" that expresses individuals' creative relations with complex systems from Bell Telephone to software (Chirillo, 2001). Tweaking, however, is focused on making adjustments at the level of human life processes. The two tweaking examples we have chosen for closer examination are the administration of fish oil products to school pupils in attempts to improve examination results and the application of principles of neurolinguistic programming (NLP) in the form of "learning styles" (Pritchard, 2008).

Both of these examples can be seen as postdigital versions of the "technical fix" mentality so long associated with the digital technologies discussed so far in this book—a sort of "biotechnical fix" if you like. Controversial as educational uses of fish oil and NLP are (Goldacre, 2009), it is not our aim here to debunk them. Rather we use them to illustrate and to begin to populate the wider imaginative field of tweaks. Our interest is in the potential that the biosocial imagination tweaking has for prefiguring educational futures and thereby channeling developments. As our examples will show, we will consider perceptions of and strategies toward educational futures at scales ranging from the individual pupil, through family dynamics, to local and national government. We will examine educational futures from the point of view of the biosocial imaginations that guide practical efforts to anticipate and to build personal and collective futures.

Having illustrated tweaking, we then place it in the context of a set of understandings of human capability and performance that envisage individual and social progress as movement from a "protohuman" condition first to "human" and then to "superhuman" states of being. This series has, in the past, proved helpful in gaining traction on futures by offering structure and support when inferences are made about how best to influence human life. As we will suggest later, such narratives of progress include those of "development" that follows a trajectory from "infant" to "adult" to "expert" (Newman & Holzman, 1997) and "hominization" from "animal" to "human" to "human 2.0" or to even "human 3.0" (Fuller, 2011). These visions of individual, social, and species progress, have been, and remain, powerful influences on the shaping of projected and resultant futures. Tweaking borrows some of its credibility and promise from them.

Tweaking deserves scrutiny because it has claims to make on educational futures. In the current economic and financial conjuncture, tweaking helps to cement a decades-old neoliberal commitment to global competition for and through human resources (Foucault, 2008). These commitments have been criticized for a close and sometimes exclusive identification of persons with life processes (Agamben, 1998). Currently, neoliberalism intersects with understandings of human and educational futures to produce a concern with the enhancement of "normal" performance (Sandel, 2006). This has consequences for the understanding and practice of "agency" by parents, pupils, and teachers as individuals and by state agencies, such that being agentic in educational settings can in some circumstances become coextensive with the individual and/or collective pursuit of improved examination performance. Since it is not clear to the authors that all possible futures will respect good grades, we finally complement our critical commentary by sketching a biosocial imaginative frame that may help foster alternative educational futures.

Biosocial Imaginations

The term "biosocial imagination" indicates that attempts are often made to understand relations between life processes and lifestyles, frequently in order to make both or either of these tractable to human plans and desires (Lee & Motzkau, 2012; Lee & Motzkau, forthcoming). Biosocial imaginations, as we conceive of them, are not flights of fancy, but are created and used to serve practical ends. Further, we do not imply a simple contrast between "facts" and "imaginations" of the kind that is often used to detect myths and fantasies that deviate from a scientific authorized version. The role of metaphor in shaping and guiding bids for tractability and the intimate connections between biosocial imaginations and practical settings mean that we are dealing here with relationships whose complexity exceeds the analytic grasp of a fact/fantasy distinction. Often, for example, an imagined relationship that appears to have delivered desirable outcomes in one area of activity will be adopted as a guiding metaphor in quite another area. A good illustration of this would be the concept of "culture." Originating at the intersection of the life processes of some plants with human attempts to increase and diversify their yield, the notion of cultivation came to be applied to humans, especially the young, in eighteenth-century Europe (Velkey, 2002).

The culture metaphor may be questionable, and its outcomes mixed—in some contexts reinforcing inequality (Bourdieu, 1986), in others encouraging effort and giving hope—but the concept of culture has, nevertheless, had considerable impact. Even though, in fact, human life processes differ a great deal from those of crop grasses, the imaginative transfer certainly did guide practical interventions in human lives. For these reasons, we take the view that fact/fantasy or fact/myth distinctions would be inadequate analytics to apply to contemporary understandings of educational futures. Further, since the future simply has not yet arrived, nobody is in a position to speak authoritatively about it. This does not mean, however, that we cannot become more articulate about existing preferences, strategies, and emergent prefigurations, hence our deployment of the analytic device "biosocial imaginations."

Recently, developments in the life sciences such as behavioral genetics (Kelsoe, 2010), pharmaceuticals that affect cognitive performance (Harris, 2010) and the advance of self-help and therapeutic traditions such as NLP (Ready & Burton, 2010) have promised to render life processes increasingly open to direct and targeted intervention. For some, the hope is that accurate knowledge of life processes will allow us to set our metaphors and narratives aside in favor of more effective techniques—imagination yielding to fact. While metaphorical models and narrative expectations could be applied to whole individuals in all their complexity and capacity to surprise, a pharmaceutical or neuroscientific intervention might promise to tweak individual cognitive performance or teaching practice to bring immediate, predictable improvement as measured against a selected standard of excellence.

It is clear to us that some tweaks may well prove effective in their own terms and that they may find complementary places alongside other metaphors and narratives in guiding educational practice. In what follows, however, we will critically examine this trend. Despite the claims and counter claims about scientific insight and authority that often accompany tweaks, we will consider tweaking as a variety of biosocial imagination. We will suggest that tweaking has the potential to set unnecessary limits to imagining what education can be and what young people might become.

Tweaking, Education, and Political Economy

In the current effervescence of interest in and capitalization of life processes, it is easy to overestimate the power and the durability of innovations that seek their authority in the life sciences (Harvey, 2010). Tweaking, however, also has a considerable affinity with contemporary governmental concerns to secure cognitive resource (Foresight Mental Capital and Wellbeing Project, 2008) that present links between human lifestyles and life processes in terms of the maximization of performance at minimal cost. Part of its contemporary attraction, as we see it, is that it fits neatly into an already persuasive conjunction of understandings of economic competition, the enhancement of human ability and social progress. Further, success in formal examinations has enjoyed the status of a dominant metric of ability and progression for many years. Tweaking, then, has the potential to reinforce tendencies to prefigure young people's lives in the relatively narrow terms of success or failure in obtaining qualifications. These connections between biosocial imaginations, the functioning of modern states, and the politics of education merit further explication.

Each time a state agency takes measure of a child's physical growth to assess it against a preferred age-related norm (Turmel, 2008), a link is established between the child's present condition, the life processes involved in her maturation, and a collective, often national, future. This link configures the child as a site of intervention at which futures can become, at least partially, tractable from the standpoint of the present. Thus, if the child's growth is suboptimal, steps can be taken to improve it on the basis of that measurement and through the application of expertise. That the young can and should be configured as human fragments of the future tends to be a core assumption of state policy. Configuring them in this way has become a core process of the practice of maintaining a state. It also provides the grounds on which a range of relations between human lifestyles and life processes are imagined and assessed. In the terms of this chapter, this configuration of the young has established the grounds for contestation between various forms of biosocial imagination in educational settings.

A markedly "Romantic" biosocial imagination, for example, might see the child as a site of natural virtue and spontaneous growth, whose burgeoning energies can only ever be harnessed or repressed by civilized social institutions. The child's destiny as an adult is either to change these institutions or to slumber within their confines. It might pose such questions as follows:

- How free is this child to make a future of her own?
- What help might she need to bring that future about?
- Is her integrity being respected?

Within the frame of another, more analytically inclined, biosocial imagination, the key questions to be asked about the young are as follows:

- What capabilities do they have?
- Which of these capabilities are innate and which can be added?
- How can they be shaped now so as to deliver future preferred outcomes?

At times, the contrasts between these imaginations have been made very clear. They have even, as in conflicts between "progressive" and "traditional" approaches to education, or between "genuine" education and "mere" training, been aligned with broader political sensibilities and socioeconomic voices and set in stark opposition to each other. Contemporary educational debate and practice tends to have space for both and, though they still inform some of the key tensions between different visions of good educational practice, they can also inform and energize each other.

In this context, a good tweak stands out because it offers a chance to undercut ideological conflicts and to bypass tensions, however creative. Consider the issue of disorderly classrooms. For a Romantic, this arises in proportion to youngsters' lack of authentic freedom, while a disciplinarian imagines a set of moral identities and social interventions swinging into action to check nascent disorder. But what if a disorderly classroom was, in part, a result of imbalances in children's diets? Could not a wisely chosen dietary supplement, such as a fish oil preparation, soothe the classroom from within the body of each pupil? Depending on circumstances, this might be seen either as enabling the transfer of discipline to more productive areas or even, where children's diets are poor, as a form of liberation. Either way, a good tweak promises to allow us to dispense with the question of how we should view classroom order. It promises an alternative to the work of educational metaphor in delivering futures. As should become clear in our examples, part of the charm of tweaking as a form of biosocial imagination is that, even as it relies on and generates imaginary relationships it is able to present itself as a simple and practical alternative to the time consuming business of debating social relationships in educational contexts.

Some of the metaphors that currently shape educational practice are economic in nature—investment, cost, and return. In this context, the main characteristic of a tweak is that, since it is highly targeted and appears to involve minimal material change, it can present an attractive cost/benefit ratio. Why change classroom practice when a dietary supplement could be effective? Why reduce class sizes when a change in classroom practice could help? Here, the tweak is reminiscent of the nudge of behavioral economics, where a small change in "decision architecture" promises large-scale behavioral change. In our view, in Anglophone regions at least, the tweak and the nudge share the political-economic context of decades of downward pressure on public spending coupled with rising awareness of the scarcity of resources of all kinds, from fuel, through water to desirable forms of paid employment. Thus, the attractions of the tweak and, arguably, of wider contemporary interest in human enhancement is underscored by the individualization of risk and opportunity (Beck, 1992) and the increasingly high national and personal costs of failure to compete effectively.

Tweak One: Fish Oil

In 2007, Durham County Council (DCC) in the northeast of England investigated the effects of a fish oil dietary supplement on pupils' GCSE exam results. They began with 3 thousand pupils at year 11 taking supplements at home and at school. The fish oil supplements were provided free of charge by the company Equazen, manufacturer of such products as the widely available "eyeq chews," a fruit flavored, sweetened preparation of "naturally sourced" Omega-3 and Omega-6 oils. By the time of the national GCSE examinations (held at the end of the school year), around eight hundred pupils were still compliant with the program. In order to estimate the effects of the supplements on GCSE outcomes, the Council's Children and Young People's Services division compared the results of children who had remained compliant with those of children who had not taken the fish oil that Equazen provided. The two groups' performance did differ, with persistent Equazen users scoring higher than the others.

This investigation generated a great deal of positive publicity for fish oil supplements. Its status as scientific research has, however, been called into question by a series of closely argued articles by the science journalist Ben Goldacre (2009). He points out that there is no good reason to attribute the differences in performance to the oil, given that the compliant group was self-selected and perhaps more invested in educational achievement than the others and that no attempt was made to control for placebo effects. Further, no information was sought about the diets and supplement use of children who did not take the Equazen product.

In a press release (DCC, 2008) the Head of Achievement for DCC's Children and Young People's Services acknowledged that the study's design did not allow any positive inference to be drawn about the effectiveness of fish oil in raising children's achievement. Having said this, however, he pointed out that had no difference been detected between groups, Durham would have been likely to dismiss fish oils entirely. Combining this imaginary negative result with the actual but scientifically meaningless positive result enabled him then to maintain hope in the effectiveness of fish oils:

> Taking all this into account, it is our view that this study has produced some interesting and possibly exciting issues for further investigation that could be the basis for future scientific trials.

While they could be argued to fall short of conventional scientific reasoning about effectiveness, we would suggest that these maneuvers allowed the local council to maintain something they valued. In other words, the possibility of a strategic alliance between Durham Children and Young People's services, Equazen, and aspirational service users was organized around a tweak that could help meet policy objectives of increased GCSE achievement. We do not suggest that there was any ethical impropriety here. Further, whether this trial was "good" or "bad" science and whether fish oils really can raise performance are important issues, but they do not concern us here. Instead, we consider the Durham case as an example of the flexing and testing of a tweak. It is interesting to note just how durable the connection between fish oil supplements and exam performance proves within this example. Some of DCC's key employees were attracted by the possibility that a small, relatively inexpensive change in children's diets could bring about an improvement in their performance that would benefit not only individual children, but also DCC's reputation with its publics and with other state agencies that monitor its performance using GCSE results as a metric of success. This attraction, based though on the supposition that fish oils act directly on children's brain chemistry to alter their cognitive functioning, is able to survive scientific scrutiny arguably by dint of its promised direct effect.

We make no claim that associations of fish products and cognition are novel. Fish products have been linked with brain function long enough for the connection to become proverbial. But the Durham case is a good example of how the imaginative form of tweaking operates. There are indications that others shared DCC's interests and assessment of possibilities. It seems likely that some parents and children, for example, saw fish oil supplements as a potential practical means of making exam performance tractable. Tweaking offered them a fresh pathway to agency in relation to children's personal futures. It also offered Equazen publicity and marketing opportunities for their products.

This use of fish oil probably seems low risk to its users, it is, after all, a repurposing of an existing foodstuff. It takes its place in a thriving market of nonprescription boosters and supplements addressing such phenomena as "immunity" and "mood" that is relatively unregulated in Europe (Coppens, Fernandes da Silva & Pettman, 2006). That said, there is also considerable interest in the use of pharmaceutical products to provide tweaks (Harris, 2010) despite the fact that the use of substances in this category is relatively tightly regulated on safety grounds. There is a clear pathway to agency that gives young people a greater sense that they have traction over their own performance and futures that involves Ritalin. Developed and prescribed for Attention Deficit Hyperactivity Disorder, this product has found its way into informal use as a cognitive enhancer. It is now used by large numbers of students in attempts to boost their ability to concentrate, especially at times of exam revision and writing to deadline (Harris, 2010). What remains to be studied here is whether and how users gauge effects, what metrics they use to decide on their dose, and how closely these forms of knowledge are shaped by supply chains, by peers, and by internet communities (Coveney, Gabe & Williams, 2011).

Tweak Two: Neurolinguistic Programming

There is insufficient space in this chapter to give the history of neurolinguistic programming (NLP) the attention it deserves but key aspects of its development have made it a rich source of marketable educational tweaks. NLP is often understood to have its origins in the work of the twentieth-century US psychiatrist Milton Erickson (1991). He developed ways of inducing trance states in his patients to effect therapeutic change. Consider his account of inducing a trance state while shaking someone's hand:

> When I begin by shaking hands, I do so normally. The "hypnotic touch" then begins when I let loose. The letting loose becomes transformed from a firm grip into a gentle touch by the thumb, a lingering drawing away of the little finger, a faint brushing of the subject's hand with the middle finger—just enough vague sensation to attract the attention. As the subject gives attention to the touch of your thumb, you shift to a touch with your little finger. As your subject's attention follows that, you shift to a touch with your middle finger and then again to the thumb... Then almost, but not quite simultaneously (to ensure separate neural recognition), you touch the undersurface of the hand (wrist) so gently that it barely suggests an upward push. This is followed by a similar utterly slight downward touch, and then I sever contact so gently that the subject does not know exactly when—and the subject's hand is left going neither up nor down, but cataleptic.

For the developers of NLP (Bandler & Grinder, 1981) accounts of this kind have been inspirational. The theory that runs parallel to this kind of practice is that human minds are so good at developing habits and expectations that the "handshake" which, in reality, is a complex series of coordinated actions, can be perceived as and, at a more fundamental level, processed within the brain as if it were a single event. Note that Erickson expresses himself in terms of events at the neurological level. This tendency of human information processing to form habitual expectations is usually useful because it keeps everyday life relatively simple. The brain already "knows," as it were, what is going to happen next from the moment the handshake is initiated, or, at least, carries very clear expectations that are usually met. Erickson's pattern of touches is intended to disrupt those expectations and, thereby, places the person in a "trance" state of alert confusion. For Erikson, once in such a state, the individual is open to therapeutic suggestion. To express this colloquially, since the brain does not know what is going to happen next, the person becomes open to being shown the way.

The general principle that has been developed into NLP and that we examine here is that, thanks to quirks of human perception and information processing, it seems possible to have a major effect on others by carefully adjusting relatively small aspects of one's own conduct. The concept of multiple learning styles—visual, auditory, and kinesthetic (VAK)—that is popular in educational practice today was developed from just this background by NLP practitioners (Pritchard, 2008). Different pupils at different times have different levels of engagement with classroom activities. Some children have a characteristically low level of engagement. Where this has been seen as problematic, a variety of means have, historically, been deployed to render their level of engagement tractable. Violence, moral suasion, classroom layout, and appeals to long-term self-interest have all had their place, among many others. Each of these has carried its own idea of what a pupil consists of that is consistent with the mode of intervention. Thus, for the violent, a pupil is a kind of incalcitrant beast. For the moralist, a pupil can behave badly, but still be shown a better way so that they don't "let themselves down." Within the terms of the VAK learning styles approach, however, the pupil needs to be engaged with as a neurological information processor.

The supposition on which the learning style approach is based is that, even though each such processor has its own characteristic tendencies and preferences, they tend to fall within one of three clusters. Visual learners find it easier to engage with information when it is presented in visual form. Auditory learners learn best through listening. Kinesthetic learners like to move about and touch things. All this is taken to be a function of individual neurological differences. If this is the case and the aim is to maximize all pupils engagement with classroom activities so that they all learn as much as they are able, then, saving the segregation of classes on the basis of learning style, it would make sense to provide a mixture of information modalities in any given lesson. Further, if all the above holds good, it may make sense to test pupils or for learners to test themselves to discover what sort of learner they are.

The concept of learning style as derived from NLP is controversial. Recent reports indicate that the evidence base for its effectiveness is weak and that the hypothesized connection between events at the classroom scale and at the neurological scale are not understood well enough to form the basis of classroom practice (Pashler, McDaniel, Rohrer & Bjork, 2008). In the terms of this chapter, however, learning style is a good example of tweaking as a form of biosocial imagination. Just as fish oil offered a range of new pathways to agency so does VAK. Teachers can use it in attempts to square the circle of passing diverse individuals more or less smoothly through the same institutional requirements of attendance, attention, and performance. So it does something to legitimate or to accommodate pupil's cognitive, attentional, and attitudinal diversity whether this is, in reality, channeled into three clear "styles" or not and regardless of the veracity of NLP claims. In this, like fish oil, it bypasses the issue of whether or not it is really effective at the level of life processes by having effects at the level of institutional processes.

Ultimately, no single tweak is immune to lack of evidence and or to sustained controversy, but the broader biosocial imagination that we have called tweaking is fairly durable because it can spread its risks. If one tweak proves a failure, another can be sought. Thus if fish oil were to fail, Ritalin or some other substance could take on its imaginative burden. Since individual tweaks promise traction on the future, for pupils, parents, teachers, and educational organizations, and because tweaking as a form of practical, biosocial imagination is relatively resistant to evidence and thrives within controversy, tweaking may well continue to influence educational practice and thinking about educational futures.

Protohuman-Human-Superhuman

As we have suggested so far, the variety of biosocial imagination we have called tweaking does not arrive out of thin air. In our view, it is a relatively recent development in a long history of attempts to engage with human life processes so as to direct change. The terms "protohuman" and "superhuman" clearly carry a lot of baggage. They resonate, for example, with racism and fascism. The term "human" covers a lot of controversies too (Fuller, 2011). But these terms, and the series they compose, also express something important about those biosocial enterprises that focus on human life processes whether they are undertaken by legitimate or illegitimate authorities: knowledges about human life processes and understandings of their tractability have an intimate relation with attempts to build power and to influence future events. Just as children are often considered human fragments of the future, so such knowledges have offered both the means to work these fragments into realized futures and the visions of future possibility that steer present activity. So we are not using these loaded terms in order to endorse them, but to highlight those intimate relations between knowledge and power, humans and futures, that are, sometimes, but certainly not always, illegitimate, ill-advised, or exploitative.

There are two imagined temporal series of the general form "protohuman-human-superhuman," that we would like to examine here so as to fill in some of the context around tweaking and futurity. One of these—development—might sound scientifically outdated. Another yet—hominization—might seem excessively speculative. Despite these drawbacks, we address them because their existence says something important about biosocial imaginations and futures. They indicate that a powerful majority of biosocial imaginations that have led, thus far, to the conceptualization and realization of futures have restricted themselves to working on humans and on human life processes. They are "anthropocentric" biosocial imaginations (Lee & Motzkau, 2012) that present humans both as the most proximate and the most significant resource for future making and as a resource that can be considered in distinction and worked on in isolation from all others. As we shall see shortly, tweaking does something rather interesting in this context.

First, consider Piaget's conception of the newborn human:

Let us imagine a being, knowing nothing of the distinction between mind and body. Such a being would be aware of his desires and feelings but his notions of self would undoubtedly be much less clear than ours. Compared with us he would experience much less the sensation of the thinking being within him, the feeling of a being independent of the external world. The knowledge that we are thinking about things severs us in fact from the actual things. (Piaget, 1927, p. 37)

Apart from anything else, Piaget's cognitive developmental psychology was a remarkable piece of biosocial imagination. Here he offers us an imaginative portrayal of an infant as a being quite different from those adults who (we must assume) are Piaget's intended audience. To become human in the way that Piaget intends, that is to say, to become a being that lives in the full recognition of its ontological isolation from the nonhuman world, the infant must develop and learn. Through the accrual and processing of experience, it needs to pass through such, for Piaget, faulty ontological commitments as the "animistic" belief that the nonhuman world is moved by nonhuman intelligences. It must forsake "magical thought," the idea that its wishes alone can bring about material change. Once it has made this passage and come to recognize its fundamental separation from actual things and to see the material world as a dead space, it is in the "rational" position, which is uniquely human, of being able to deliberately intervene in the nonhuman realm reliably to produce desired effects. These ontological commitments were contested in even Piaget's time (Whitehead, 1927). But Piaget had great influence in education and elsewhere.

To Piaget's series "infant-child-adult," we take the liberty of adding expert. We have been encouraged to do so by Newman and Holzman (1997) who point out that within Piagetian rationalism, those adults who are most surely and closely acquainted with their separateness, most skilled in the forms of detached reason that approximate "objectivity" are in a position to use that separateness as the firm ground from which to deliberately intervene in human development and behavior. So we have a series of the kind protohuman-human-superhuman in which the infant, the adult, and the expert take the respective roles in a sequence that speaks to the passage of time and the correct location of power to shape humans and, thus, futures. As the adult guides the child, so the expert guides them all.

Second, consider hominization, the evolutionary passage between animal and human. Much contemporary paleoanthropology (Stringer, 2011) indicates that the evolutionary processes that distinguished the life-forms we call human from other closely related species involved multiple protohumans and human species, who migrated, competed, met, and interbred time and again. Human evolution was likely not a simple progression from a single homogenous protohuman species to a homogenous human species. Nevertheless, the question of how nonhuman animals turned into human animals has often been presented in the form of such a temporal series—a progression from past into present, from protohuman to fully human. This narrative structure has recently been embellished by discussion about the impact of digital and other technologies on the status "human." For some, technologies are bringing about a new evolutionary shift from human to human 2.0 (Fuller, 2011). In these views, humanity is now changing just like protohumans once did. This time, however, change is taking place on the temporal scale of human generations. Others, yet, argue that we are now seeing human 3.0 because the development of literacy already took us to 2.0 (Changizi, 2012). These issues play a large part in structuring educational discourse concerning "digital natives" (Prensky, 2010) and educational technology more generally.

It is essential for present purposes to note the parallels between the sequences that have been imagined for individual human development and for humanity's evolution as a biological specie. These parallels between how the specific and the general case are imagined helps to produce a sense of total coverage of biosocial possibility. As we have argued elsewhere, this foreclosure of possibility is a characteristic of anthropocentric biosocial imaginations. In both general and specific cases, these sequences organize pasts in such a way as to suggest clear, navigable paths toward the future. Where these sequences have guided the anticipation of the future, the tendency has been for futures to become the work of a relative few that is restricted to the manipulation of humans and their cognitive abilities.

The two tweaks we examined above certainly offer the superhuman position of expertise. They promise educators the tools to allow them to occupy that position and thus to play a part in shaping the future. But tweaking does more than this. In an enabling cultural context of high cultural prestige for the life sciences, tweaking draws both on the "technological sublime" (Nye, 1996) and on the traditional figure of the expert to offer agentic positions from which it seems possible to become an expert over oneself at the level of one's own biochemistry or neurological habits so as to effect step changes in your own abilities (Rose, 2007). In a political economy concerned with competition for scarce resources, including the resource of desirable employment, and where that competition takes place in large part through public examinations, tweaking would appear to have a strong grip on educational futures.

Developmental and evolutionary series, just like the two tweaks we examine above certainly can be addressed as if they were simply sets of propositions the validity of which can be tested. It is always possible to address them as if their significance were limited to whether they are either factual or mythical. Our strategy, however, has been to read them as structures of practical imagination, as attempts, however questionable, to organize pasts so as to provide, at the very least, reassurance about the tractability of the future.

Infrahuman-Human-Ultrahuman

Links between human life processes and power relations among humans are often contested in terms of their factual basis. Our perspective yields a fresh observation, however. Dominant forms of biosocial imagination that affect educational debates focus more or less exclusively on specifically human life processes and the traction they offer on the future. These are anthropocentric in character. In our tweaking examples, there are elements of human biochemistry and there are human neurons, but there are no bacteria and no carbon cycle. This anthropocentrism is more tightly drawn as tweaking apparently brings human plans and desires into more immediate relation with life processes than ever before. Tweaking thus helps to enact the separation of human self-comprehension from comprehension of wider life processes that Piaget endorsed. We find the relations that tweaking establishes between life processes, human agency, and futures worrying because, as sustainability issues come to the fore, versions of the future that are predicated on the separateness of human from other life processes look ever less likely to be viable. This leads us to ask what frames other than anthropocentrism might exist, and what potential they might have to generate fresh biosocial imaginations. We think it is important to find positive alternatives here because of the risk that critical voices become simply reactionary, attempting to dismiss the significance of the life sciences for educational futures.

We noted earlier that during the eighteenth century the cultivation of crops became a rich source of metaphors that came to guide educational practices focusing on the cultivation of children, so as to build or to preserve national cultures. Though crop science and issues of food security are still important today, they have been joined by fresh concerns and fresh images of relations between humans and the nonhuman living world. Some of these contradict the image of human separateness and distinctiveness that Piaget endorsed, which has often guided perception of human evolution, and that resonate in tweaking. For the biologist Margulis (Margulis & Sagan, 1997), human life on earth is dependent not only on the lives of multicellular organisms like animals and plants, but, most fundamentally, on bacterial life. If, for example, the bacteria involved in the decomposition of dead multicellular organisms were suddenly removed, an insuperable disposal problem would rapidly ensue. Further, these relations of dependency exist between not only human societies and bacteria considered as features of the external environment, but also within human bodies. For example, a host of commensal bacteria aid human digestion and immunity. For Lovelock (2000), human activity is just one factor in a wider earth-wide system that links the albedo of the earth's surface, the cycling of water, and of atmospheric gases through the life and death of life-forms and through the formation and destruction of geological deposits of ice, limestone, oil, and gas. Within limits, the system he calls Gaia, is capable of homeostatic regulation of temperature and atmospheric composition. For many observers of climate change (IPCC, 2007) it has become apparent that human lifestyles are capable of stimulating homeostatic correction against immediate human interests.

The series Protohuman-Human-Superhuman presents a single line of development that consistently isolates humans over time. It presents the future as change with direction that can be set by expert human intervention. For us, taking Margulis and Lovelock into account results in a rather different series: Infrahuman-Human-Ultrahuman. Our term "Infrahuman" refers to the bacterial and eukaryotic cell activity, including neurological activity, that composes human life and links it fundamentally with other life processes. In our usage, "Human" does not refer primarily to a distinctive biological specie but rather to those levels of institutional life in which we find ourselves as individuals with autobiographies who depend upon and create social organizations. Our term "Ultrahuman" refers to the large-scale transformations that detain Lovelock and to our involvement as individuals and as members of social organization in them. Just like Protohuman-Human-Superhuman, this frame brings the biological and the social closely together. However, rather than presenting the past and the future as an exclusively human business with a clear direction, it presents a constantly shifting set of relations between human activity and other life processes.

Even as we present this alternative frame, however, we would caution against attempts to read it as akin to a manifesto intended to drive change. In our view, the metaphorical transfers and imaginative associations that have, in the past, enabled people to navigate toward futures have rarely offered firm and consistent ground. Biosocial imaginations are better understood as forms of curiosity and as generators of questions that act as sources of programmed change. Thus if the frame we present is to alter the enactment of education to make tweaks less attractive, it will do so in so far as it is seen as helpful in enabling practical responses to wider change such as the effects that increasing resource scarcity may have on the viability of continuous economic growth.

Conclusion

In this chapter we first introduced the concept of biosocial imagination considered as a feature of the organization of educational effort and illustrated its relation to educational futures. Using two examples of tweaking, we then identified a contemporary style of biosocial imagination and indicated the means by which it can shape educational futures. Our criticisms of tweaking were first, that it narrows thinking about what individuals can become through education by channeling agency toward, for example, the improvement of exam results and, second, that it presents a narrowly anthropocentric view of the potential significance of biosocial relations for educational futures.

Drawing on contemporary biosocial concerns about relations between infrahuman, human, and ultrahuman phenomena, we then proposed a generative frame that may play a part in the emergence of fresh biosocial imaginations that are alternative to tweaking. It is not our intention here to anticipate practical developments. Rather we are drawing attention to a possible source of future biosocial imaginations that could exploit the promise of the life sciences for wider ends than anthropocentric forms like tweaking currently do. As we see it, this would not depend on individuals gaining full and certain knowledge of the very complicated relations that Margulis and Lovelock highlight. Rather, as in the case of the metaphorical transfer of culture from agriculture to education, strategic simplification is the principle virtue of biosocial imagination as a resource for reframing educational activity.

CHAPTER 11

Epilogue: Building Allegiances and Moving Forward

Keri Facer and Neil Selwyn

Introduction

The chapters in this book have certainly resulted in a markedly different portrayal of education and technology than is usually the case with academic discussions on the topic. None of these chapters has concurred with the notion that educational technology is simply a matter of "harnessing" the "power" of largely neutral, benign technologies in ways that support and empower individual learners. Instead, as all the authors in this collection have shown, there is a lot more to education and technology than simplistic descriptions of "technology-enhanced learning" and "personalized learning networks" suggest. Indeed, the important point of contesting the common-sense language of technology and learning was ably critiqued in the chapters by Friesen and Williamson, and remained a recurring theme throughout other contributions. The plurality of interests, values, and agendas that have come to inhabit seemingly unassailable notions such as "twenty-first-century skills," "e-safety," and "one-to-one computing" was laid bare in the contributions from Shutkin and Hope. The chapters by Rudd and by Dessel, Ferrante, and Sefton-Green remind us of the influence of national politics and the ambitions of nation-states, while Selwyn also reflects the importance of interests operating above/below the level of the nation-state.

Finally, the chapters from Hall and from Lee and Motzkau stress the importance of broadening our imaginations beyond the digital artifact and toward the issues of ethics, exploitation, and the environment and the postdigital turn to biological, neurological, and cognitive technologies. As such, these last two chapters offer a glimpse of where this field is heading—outlining the significance of the postdigital in twenty-first-century education. These latter chapters should certainly prompt us to begin to consider what the *next* wave of technological change may mean for the critical study of education. This implies an increased focus on what Steve Fuller (2011) calls the "converging technologies" of the nano-, info-, and cogno-technosciences as well as biomedical and genetic engineering and the so-called green technologies. If we can be sure of one thing, it is that the education of the twenty-first century will not be influenced and shaped only by our present technologies and practices.

Issues of future critique notwithstanding, the main aim of this brief final chapter is to make some sense of what has emerged from the preceding ten chapters. Paying attention to the politics of education and technology, as all these chapters have done, brings with it some risks. Most obvious among these is the accusation that critical scholars simply sit on the sidelines pointing out mistakes while other people roll up their sleeves and actually get things done in educational settings around the world. However, this accusation can be countered by the evidence in these chapters that action *without* critical reflection on the politics of the situation is almost always counterproductive. Understanding what actually happens in the struggles over imagining and implementing technologies in education is essential if the potential mobilization of educational technologies for social justice and equity in education is to stand a realistic chance of being achieved.

As a whole, then, the chapters in this book offer good grounds for understanding educational technology as profoundly political in nature and form. The wide-ranging forms of "technology" reviewed over the past ten chapters are clearly functioning as sites for the "working out" of various educational struggles and conflicts—giving symbolic form to dominant values, interests, and agendas as well as grounding them in new sets of educational processes, practices, and relationships. In these cases, educational technologies act as a powerful cultural support system to wider dominant agendas in education. There is little value, therefore, in the pursuit of specific debates over the "rights" or "wrongs" of particular technologies in education (e.g., the value of providing an iPad for every student, or the mass provision of free higher education through Massively Open Online Courses (MOOCs)). These are all proxy battles for much wider controversies and conflicts surrounding the nature, form, and function of education in the twenty-first century—that is, tensions between market and state, private interests and public good, and the primacy of individuals as opposed to the collective.

Of course, it is important to acknowledge that the relationship between educational technology and these wider conflicts is not wholly deterministic. It would be wrong to conclude that digital technologies are simply direct materializations of wider ideological values and agendas. Instead, the political nature of these educational technologies needs to be understood in more nuanced ways. Indeed, implicit throughout our discussions has been the notion that the politics of education and technology is not always easy to identify or discern. Of course, the fact that it is difficult to do so is itself an indication of the ideological form of educational technologies. The general sense of new technologies being inherently "beneficial" is a dominant position within the field of education, and therefore a difficult orthodoxy to challenge.

There are many influential interest groups acting to promote dominant values as commonsensical and to maintain what Freeden (2003, p. 108) terms "the impenetrable and non-transparent shield of self-evidence." That said, the past ten chapters have done much to deconstruct and look beyond these technologies' fuzziness of definition and apparent multifaceted appeal to interrogate the various interests and agendas that they come to serve. In particular, an obvious conclusion to draw is that despite their cutting-edge, disruptive, and innovative connotations, the underlying political function of these technologies is often one of adjusting educational provision and practice to a set of underpinning conditions where primacy of individual action and market forces is aligned with a set of skills, behaviors, and dispositions linked with the demands of the new economy. Indeed, in many ways it is difficult to look beyond the influence on current forms of educational technology of the overwhelming imperative of twenty-first-century capitalism and the demands of the "new" economy. The digital forms of technology-based education that have featured throughout this book have clear links with the needs of immaterial labor—that is, self-directing, self-disciplined, and routinized workers who are comfortable working with (and within) informatic and algorithmic environments. In this sense, digital education could be seen as little more than the preemptive organization and exploitation of future immaterial workforces.

In this sense, the politics of education and technology could be seen as nothing new per se. It could be argued that beyond superficial differences in physical form and appearance, the forms of new technology considered in this book represent simply the latest material means of the production and maintenance of the capitalist social order. Thus while the exact characteristics may have changed, the dominant frames of meaning of these forms of education continue on from those that they replace—that is, maintaining arrangements that benefit elite interests at the expense of the majority. As such the allegedly new features of contemporary education could be said to relate strongly to the well-established logics of accumulation, not least the overwhelming imperative of capital and the maintenance of the rate of profit.

That said, in recognizing these continuities we should not overlook the potential differences and novelties of these recent forms of technology-led education. For example, while old models of power appear to remain intact, it would appear that there are a number of new forms of power relations attached to the increased use of these technologies—what Matthewman (2011, p. 60) describes as, "new practices, new observations, new organizations and new knowledge." Certainly, the technologies outlined in the past ten chapters appear to herald a dematerialization of ideology, marking not only the intensification but also the further displacement and normalization of values of individualization, marketization, and new capitalism in educational contexts. As such, there is much here to challenge and critique—not least in terms of exploring further the obvious contradictions and cracks within the apparent hegemony of educational technology.

To be content with a position of disengaged critique alone, however, strikes us as increasingly unsatisfactory. Indeed, when we examine the history of social and civil movements that have successfully challenged the dominance of elite power in education, it is clear that critique, while necessary, is far from sufficient for effecting meaningful social change. Rather, if we are concerned with the creation of modes of education that successfully resist the colonization of the lifeworld by the market, the role of the researcher must be (following Raymond Williams) to "make hope possible" by presenting alternative practices and trajectories (Williams, 1958). Indeed, the ultimate purpose of problematizing what is being presented currently to us as "educational technology" is to give serious thought to how we create alternative, fairer ways of using digital technologies in education.

And it is here, perhaps, that the small body of previous critical analyses of technology in education has been at its weakest. Even if they could be considered to have successfully documented the patterns of power and inequality that characterize the introduction of new technologies in education, most such studies (our own included) can be rightly criticized as failing to systematically construct alternative trajectories. In a context in which educational change is being driven primarily by design disciplines (in other words by those who are actively engaged in experimenting and constructing alternatives to what currently is being offered), a sociological stance characterized primarily by description and critique can only ever have limited purchase in constructing more equitable alternatives. Equally, in a context in which many advocates of technological change in education are themselves inspired by aspirations toward more equitable and democratic education practices, a disciplinary tradition that is characterized by often-disengaged critique rather than committed collaboration to build alternatives, will be treated understandably with suspicion and skepticism.

In this epilogue, therefore we want to reflect briefly on strategies for *engaged* critical research in education and technology. In so doing, we want to uncover resources for "non-stupid optimism"—a term that is drawn from Erica McWilliam's (1998) work in schools, which reflects the necessary balance between skepticism and hope that we feel is particularly relevant for the study and design of education and technology. In particular, we want to reflect on the thorny question of how academic writing and research in this field is imagined, funded, and conducted. In so doing, we want to make the case that there are currently underused methodological traditions that are capable of building genuinely useful knowledge and making a contribution toward greater educational justice.

Engaged Critical Research in the Politics of Education and Technology

The first, and probably most important, step toward engaged critical work in this field is to recognize the limitations of research that is wholly directed and conducted within the academy and wholly directed and conducted within the agendas of the mainstream educational technology arena. This reflects the need for academics to recognize reflexively the extent to which universities themselves are sites through which existing power relations are maintained and justified, and through which political struggles are conducted (Bourdieu, 1988b). This collection of chapters, for example, and despite our best intentions, represents the accounts in the main Western, white, able-bodied, and predominantly male academics in full-time employment. Absent from the discussions here are the insights and experiences, the knowledge and expertise of the precariously employed, of working class women, of many from the global South, and of those who are most marginalized by current education policies as well as those who are working on the ground in schools and colleges. As well as being a perennial issue in English language academic publishing, this balance of inputs reflects, to some extent, the existing skewed makeup of universities in general. In UK universities, for example, 56 percent of academics (81% of professors) are male, 87 percent are white, and there are no official figures on disability or class background (HESA, 2012). Even our own attempts, therefore, make clear the need to look beyond the academy for resources of insight, expertise, and solidarity if research and writing is not merely to perpetuate existing power relations.

Any new critical study of the politics of education and technology, therefore, needs to bring its imagination and insights into ongoing and extensive dialogue with students, educators, developers, and civil society groups who are working to explore how society and education might be arranged differently. This implies an approach that encourages, in Burawoy's (2005) terms, an avowedly *public* study of education and technology, committed to the defense of the social and of humanity, which moves "from interpretation to engagement, from theory to practice, from the academy to its publics" (p. 324). This is not to say that analysis of the sort presented in this book is without merit. Indeed, such a public approach would continue to recognize one of the core strengths of academic sociology in its ability to express a "defiant" rather than "compliant" imagination, and therefore engage in the production of "really useful" knowledge that is capable of enabling radical social change (Kenway & Fahey, 2008; Boden & Epstein, 2011).

Such defiance, however, cannot continue to be mobilized as it too often is, into a rearguard defense of the rights to the autonomy of public education institutions (and those who work within them) from neoliberal policies. Instead, such defiance might be more positively mobilized in allegiance with those social movements and community organizations that are seeking also to defend the civil rights, the humanity, and the rights to education of the most marginalized in society. Indeed, the time may be particularly ripe for critically minded academics to be forming such allegiances. As Hugh Lauder and colleagues have argued with regard to a more defiant sociology of education,

> We live at a time of economic crisis when the possibilities for fundamental change are at least notionally open, and therefore it is a crucial juncture for voices within sociology of education to be heard. (Lauder et al., 2009, p. 13)

How then, might an engaged study of education and technology build such "really useful knowledge" and begin to make a difference to the nature of education in the digital age? For us, one of the most interesting sites of possibility is the growing field of collaborative research methods, also sometimes known as coproduction and codesign research (Ostrom, 1996; Fischer, 2000; Collins & Evans, 2002; Durose, 2012). Coproduced research methods seek to actively collaborate with economically and socially marginalized groups on an equal basis, recognizing the different and distinctive expertise and agendas that they will bring not only to inform, but also to shape research agendas, identify shared research questions, agree research methods, and inform political agendas. This coproduced research should neither be confused with a return to practitioner-led action research intended to improve teachers' reflection on practice and enhance adoption of technologies, familiar from many educational technology programs (e.g., Sutherland 2004; Altrichter, Feldman, Posch & Somekh, 2007). Nor should it be equated with the limited notion of "user engagement" popular within Human Computer Interaction, which also has a long track record in the field of educational technology but which has often been primarily concerned with avoiding later costly mistakes, with ensuring good interface design, and with understanding "user needs" (Druin, 2002). Rather, coproduced research is concerned fundamentally with building critical dialogue between academic researchers and community organizations around educational purpose, educational justice, and the development of strategies to achieve it. In essence, then, the aim here is to create new articulations of university and civil society to generate the sorts of strategic knowledge that can begin to support meaningful material and political change (Appadurai, 2001).

In chapter 1, we suggested that there were five steps toward a critical study of the politics of education and technology. Briefly, these were as follows:

1. Getting beyond a means-end way of thinking to interrogate how educational problems and purposes are being defined as a warrant for technological change.
2. Understanding technologies as social artifacts without intrinsic meanings, produced in social and political conditions.
3. Developing a concern with understanding the messy realities of actual technology use and implementation in education.
4. Paying attention to who benefits from technology-related policies and practices in education.
5. Maintaining a commitment to opening up new spaces for debate with a wide range of social actors about the practices and policies of educational technology.

A coproduced research agenda in the education and technology field would therefore operationalize these steps in the following ways:

1. Building allegiances with civic organizations, professional groups, and student groups who are already involved in contesting and debating educational purpose or challenging educational policy from a position of prior marginalization.

2. Working with such groups to examine discourses of technology in relation to their areas of interest in education, to examine how design and delivery decisions are being made, and, with designers and developers, to explore the potential for making different decisions in the creation of new technologies.
3. Collaborating with these groups to construct practical experiments.
4. Examining with them, how they and other groups are impacted by such implementations.
5. Working with them to build the case for social change based on this research, through the creation of new tools, public debates, publications, or events.

Examples of such engaged critical research in the field of education and technology are hard to find, although there are more examples elsewhere in education, and very few of the projects that do exist would fit into the neat categorization described above. In many cases, moreover, the outcomes of these activities are realized not in academic papers, but in the design and construction of new tools and practices, making them difficult to track down through conventional academic routes. Just looking at work in our own localities, however, it is possible to see examples emerging that begin to give shape to these ideas.

One such project, for example, is the University of Local Knowledge,[1] a collaboration between Knowle West Media Centre (a community digital arts organization working in one of the most economically excluded wards in the United Kingdom) and the computing and creative arts departments of its local universities, and other arts organizations. The project, led by the community organization, aims to challenge deficit accounts of the local area by making visible the distinctive knowledge and expertise of its people. It draws on traditions of democratic and community education to capture and share the highly diverse range of skills of this usually overlooked population, ranging from photographing royalty to caring for urban horses. Such practices are beginning to lead to new relationships with local education institutions. For example, the university veterinary schools are beginning to pay attention to the distinctive veterinary knowledge held in the area, while the local history organizations are realizing the uncovered history and insights of people living in what was once a flagship housing estate.

Another project is "80by18,"[2] led by one of the authors of this collection, which is seeking to challenge the privatization of responsibility for educational success to individual families and children by engaging civil society groups, arts and heritage institutions, charities, universities, businesses, and social enterprises in a citywide digital project. By creating a website designed to "showcase" the resources that the city has to offer its young people, from fishponds to high-end robotics, the project is designed to trigger popular debate about educational equity, collective responsibility, and the use of digital technologies as a mechanism for promoting and engaging multiple publics in educational practice. In this case, digital technology is used as a tool to make visible patterns of exclusion and inequity as well as to showcase how such patterns might be challenged.

A more internationally publicized project was Virginia Eubanks's (2011) "popular technology" approach to technology-based learning, supporting the development of technological citizenship and other forms of social movement. Working with groups of women from within the YWCA community in New York state, Eubanks has used digital technologies for a range of emancipatory projects and the development of a "popular technology project" framework for critical participatory research that is now beginning to be adopted around the world. Work here included the development of online "community asset banks" for the sharing and exchange of neighborhood skills, simulations of poverty welfare and other social service systems, the development of game-based software to support understanding of economic justice, and "community technology laboratories" supporting the cocreation of community resources.

Given these examples, readers of this book will doubtless be able to point to other such projects in their own cities that are happening outside or at the margins of university research. What is clear in these sorts of projects, however, is that the technological is not at the forefront of attention. Instead, it is the attempt to disrupt patterns of educational and social inequality that constitutes the driving force for these activities. And it is *here* that the academic researcher should be able to offer a distinctive contribution: by bringing critical and reflexive attention to the entanglements of power with which digital technologies are implicated, and by bringing the empirical research skills to examine how such entanglements are resisted or enacted through these projects. In these ways, academic research is able to usefully assist civil partners to contribute to a wider body of knowledge about how such projects work in practice. The role of the academic researcher, in these instances, is to act as "critical secretary" (Apple, Au & Gandin, 2009) and to assist projects initiated on the ground to move from individualized actions to the basis for social movements.

Such models of coproduced research, while still rare in universities and almost entirely absent in educational technology disciplines, are beginning to be supported by some national research programs. The Canadian Social Sciences and Humanities Research Council (SSHRC) University Community Partnership Program, for example, sought to promote "equal partnership between organizations from the community and one or more postsecondary institutions," with a view to creating "new knowledge in areas of importance for the social, cultural, or economic development of Canadian communities." The United Kingdom's Arts and Humanities Research Council (AHRC) Connected Communities program, similarly, is funding over two hundred projects, many of which have strong university/community collaborations at their heart, in areas ranging from social history, to environmental policy, to local politics and the use of digital technologies. This may well be an idea whose time has finally come.

Conclusions

Of course, the danger in seeking to open up research agenda setting to community partners is that such programs will be used as a way of merely co-opting research ever more closely to dominant economic interests. It could be that these good intentions result in little more than further colonization of the academy. The challenge, therefore, is for researchers to take an active role in locating themselves as part of wider movements of resistance alongside those teachers, student groups, civil society, and nongovernmental organizations who are making the case for education as a means of personal and social emancipation. And there are many such movements and groups with whom academics might ally themselves. The international student movements such as the Knowledge Liberation Front and the EduFactory Collective, for example, can teach many academics a trick or two about the reappropriation of the digital for emancipatory educational and political purposes. Similarly, student unions in all countries continue to actively seek ideas and evidence to support creative campaigning and action. Professional educators and parents, such as the New York Collective of Radical Educators are increasingly, through campaigning and research, making a powerful case for equitable education. In India, the work of the "barefoot colleges," in Brazil, the work in Porto Allegre, and across the world, the cooperative movement, are all combining education with new economic models for self-reliance and independence from the market, and beginning to mobilize digital technologies in support of these agendas.

Such organizations are often operating with constrained resources, limited research expertise, and a focus on addressing immediate concerns. Here, the critical engaged researcher can make a powerful contribution by bringing her research skills, networks, and insight into existing evidence. In combination, such partnerships have the potential to unsettle the ways in which knowledge about educational purpose, digital technologies, and pathways to social change are produced. For the university-bound academic, collaborative partnerships with community and civil groups immediately offer access to different ways of thinking and imagining the world, access to different modes of action and engagement, powerful resources for optimism. As such, changes in research practice become more widespread and embedded, they begin to construct a new "way of life" for researchers and community partners alike, a way of life that disrupts the politics of knowledge production itself. If we are looking for fertile ground to nurture the green shoots of alternative politics of education and technology, then we would conclude by suggesting that as researchers, we now need to reexamine what is taking place (or, more accurately, what is *not* taking place) in our own backyards. In other words, we now need to reshape our understanding of what it really means to be an engaged academic in the field of education and technology.

Notes

1. "University of Local Knowledge": http://www.kwmc.org.uk/index.php?project =48 (Accessed January 2013).
2. The "80by18" project can be found on Twitter, @80by18; and also: www .bristol80by18.org.uk/ [Accessed May 2013].

Bibliography

Agamben, G. 1998. *Homo sacer: Sovereign power and bare life*. Stanford, CA: Stanford University Press.

Agis, E., Cañete, C., and Panigo, D. 2010. *El impacto de la Asignación Universal por Hijo en Argentina*. www.ceil-piette.gov.ar/docpub/documentos/AUH_en_Argentina.pdf.

Allen, L. 2012. Using a cultural lens to critique the One Laptop per Child Program. In *Proceedings of Global TIME 2012* (pp. 207–212). Chesapeake, VA: Association for the Advancement of Computing in Education.

Altrichter H., Feldman, A., Posch, P., and Somekh, B. 2007. *Teachers investigate their work: An introduction to action research across the professions* (2nd ed.). London and New York: Routledge.

Amin, A., and Thrift, N. 2005. What's left? Just the future. *Antipode, 37*, 220–238.

Ananny, M., and Winters, N. 2007. Designing for development, understanding One Laptop per Child in its historical context. In *Proceedings of the IEEE/ACM International Conference on Information and Communication Technologies and Development* (pp. 107–118). Bangalore.

APA [American Psychological Association]. 2010. *Glossary of psychological terms*. http://www.apa.org/research/action/glossary.aspx.

Appadurai, A. 2001. Deep democracy: Urban governmentality and the horizon of politics. *Environment & Urbanization, 13*(2), 23–44.

Apple, M. 2000. *Official knowledge: Democratic education in a conservative age* (2nd ed.). New York: Routledge.

Apple, M. 2004. Are we wasting money on computers in schools? *Educational Policy, 18*(3), 513–522.

Apple, M. 2010. *Global crises, social justice, and education*. London: Routledge.

Apple, M., Au, W., and Gandin, L. 2009. *The Routledge international handbook of education*. New York: Routledge.

Bacon, N., Brophy, M., Mguni, N., Mulgan, G., and Shandro, A. 2010. *The state of happiness: Can public policy shape people's wellbeing and resilience?* London: Young Foundation.

Bajak, F. 2012. Peru's ambitious laptop program gets mixed grades. *The Guardian*, July 3.

Baker, L. 2001. Folk psychology. In *MIT Encyclopedia of cognitive science.* Cambridge MA. MIT Press.

Balanskat, A., Blamire, R., and Kefala, S. 2006. *The ICT impact report: A review of studies of ICT impact on schools in Europe.* European Schoolnet, European Commission, December 11, 2006. [Accessed February 11, 2012] http://ec.europa.eu/education/pdf/doc254_en.pdf.

Ball, S. 1987. *The micro-politics of the school: Toward a theory of school organization.* London: Routledge.

Ball, S. 1994. *Education reform: A critical and post-structural approach.* Buckingham: Open University Press.

Ball, S. 2000. Performativities and fabrications in the education economy: Toward the performative society? *Australian Educational Researcher, 27*(2), 1–23.

Ball, S. 2007. *Education plc: Private sector participation in public sector education.* London: Routledge.

Ball, S. 2008. *The education debate.* Bristol: Policy Press.

Ball, S. 2012. *Global Education Inc.: New policy networks and the neo-liberal imaginary.* London: Routledge.

Bandler, R., and Grinder, J. 1981. *Frogs into princes: Neurolinguistic programming.* Moab, UT: Real People Press.

Banjali, S., Cranmer, S., and Perrotta, C. 2010. *Expert perspectives on creativity and innovation in European schools and teacher training.* Luxembourg: Publications Office of the European Union, European Communities.

Bañuls, G. 2011. *Una Laptop por Niño/ OLPC en el espacio áulico. Inclusión de la conectividad a las prácticas educativas. Procesos de subjetivación en docentes y estudiantes, un estudio de caso.* Facultad de Psicología, Universidad de la República, Montevideo.

Barry, A. 2001. *Political machines: Governing a technological society.* London: Athlone Press.

Barthes, R. 1968. *Elements of semiology.* New York: Hill and Wang.

Bauman, Z. 2007. *Liquid times.* Cambridge, UK: Polity.

BBC News. 2010. India unveils prototype for $35 touch-screen computer. *BBC News Online,* July 23.

Beck, U. 1992. *Risk society: Toward a new modernity.* London: Sage.

Beck, U. 2006. *The cosmopolitan vision.* Trans. C. Cronin. Cambridge, UK: Polity.

Becta. 2004. *A review of the research literature on barriers to the uptake of ICT by teachers.* Coventry, UK: Becta.

Becta. 2005. *E-safety: Developing whole-school policies to support effective practice.* Coventry, UK: Becta.

Becta. 2006a. *Safeguarding children in a digital world: Developing a strategic approach to e-safety.* Coventry, UK: Becta.

Becta. 2006b. *ICT self review framework.* Coventry, UK: Becta.

Becta. 2007a. *Signposts to safety: Teaching e-safety at key stages 1 and 2.* Coventry, UK: Becta.

Becta. 2007b. *Signposts to safety: Teaching e-safety at key stages 3 and 4.* Coventry, UK: Becta.

Becta. 2008a. *Safeguarding children in a digital world: Developing an LSCB e-safety strategy*. Coventry, UK: Becta.

Becta. 2008b. *The state of play: E-maturity, a progress report*. Coventry, UK: Becta. [Accessed March 3, 2012] http://www.teachfind.com/becta/becta-research-research-becta-research-harnessing-technology-schools-survey-2009.

Becta. 2008c. *Harnessing technology: Next generation learning. 2008–14. A summary*. Coventry, UK: Becta.

Becta. 2009a. *Acceptable Use Policies in context: Establishing safe and responsible online behaviors*. Coventry, UK: Becta.

Becta. 2009b. *Harnessing technology review 2009: The role of technology in education and skills*. Coventry, UK: Becta.

Beer, D., and Burrows, R. 2007. Sociology and, of and in web 2.0, some initial considerations. *Sociological Research Online*, *12*(5). www.socresonline.org.uk/12/5/17.html.

Benn, M. 2011. *School wars: The battle for Britain's education*. London: Verso.

Bentley, T., and Gillinson, S. 2007. *A D&R system for education*. London: Innovation Unit.

Benyon, J., and Mackay, H. 1989. Information technology into education: Toward a critical perspective. *Journal of Education Policy*, *4*(3), 245–257.

Benyon, J., and Mackay, H. (eds.). 1993. *Computers into the classroom: More questions than answers*. Philadelphia, PA: Falmer Press.

Bereiter, C. 2002. *Education and mind in the knowledge age*. Mahwah, NJ: Lawrence Erlbaum.

Bernstein, B. 2000. *Pedagogy, symbolic control and identity* (2nd ed.). Oxford, UK: Rowman and Littlefield.

BESA. 2010. *ICT in UK state schools. 2010 summary report*. London: BESA. [Accessed March 3, 2012] http://www.besa.org.uk/documents/1467/.

Biesta, G. 2006. *Beyond learning: Democratic education for a human future*. Boulder, CO: Paradigm Publishers.

Biesta, G. 2009. Good education in an age of measurement: On the need to reconnect with the question of purpose in education. *Educational Assessment, Evaluation and Accountability*, *21*, 33–46.

Bijker, W. 1995. *Of bicycles, bakelites and bulbs*. Cambridge, MA: MIT Press.

Bijker, W., Hughes, T., and Pinch, T. (eds.). 1987. *The social construction of technological systems*. Cambridge, MA: MIT Press.

Bijker, W., and Law, J. 1992. *Shaping technology/building society: Studies in sociotechnical change*. Cambridge, MA: MIT Press.

Blaug, M. 1987. Rate of return on investment in Great Britain. In M. Blaug (ed.), *The economics of education and the education of an economist*. New York: New York University Press (pp.114–137).

Boody, R. 2001. On the relationships of education and technology. In R. Muffoletto (ed.), *Education and technology: Critical and reflective practices*. Cresskill, NJ: Hampton Press (pp.5–22).

Boden, R. and Epstein, D. 2011. A flat earth society? *The Sociological Review*, 59, 3, pp.467–495.

Bourdieu, P. 1977. *Outline of a theory of practice.* Cambridge, UK: Cambridge University Press.

Bourdieu, P. 1980. *The logic of practice.* Stanford: Stanford University Press.

Bourdieu, P. 1986. The forms of capital. In J. Richardson (ed.), *Handbook of theory and research in the sociology of education* (pp. 241–258). Westport CT: Greenwood Press.

Bourdieu, P. 1988a. *Language and symbolic power.* Cambridge, UK: Polity Press.

Bourdieu, P. 1988b. *Homo academicus.* Stanford, CA: Stanford University Press.

Bourdieu, P. 1990. *The logic of practice.* Stanford, CA: Stanford University Press.

Bourdieu, P. 1993. *The field of cultural production.* New York: Columbia University Press.

Bourdieu, P. 1998a. *Practical reason: On theory of action.* Stanford, CA: Stanford University Press.

Bourdieu, P. 1998b. *Acts of resistance: Against the new myths of our time.* Cambridge, UK: Polity Press.

Bourdieu, P. 2000. *Weight of the world: Social suffering in contemporary society.* Stanford, CA: Stanford University Press.

Bourdieu, P. 2003. *Firing back: Against the tyranny of the market.* New York: The New Press.

Bourdieu, P., and Passeron, J. 1977. *Reproduction in education, society and culture.* London: Sage.

Bourdieu, P., and Wacquant, L. 1992. *An invitation to reflexive sociology.* Chicago: Chicago University Press.

Bourdieu, P., and Wacquant, L. 2000. Neoliberal newspeak: Notes on the new planetary vulgate. *Radical Philosophy, 108.*

Boyd, D. 2011. Social network sites as networked publics: Affordances, dynamics, and implications. In Z. Papacharissi (ed.), *A networked self: Identity, community, and culture on social network sites.* New York: Routledge. (pp. 39–58)

Boyd, D., and Ellison, N. 2007. Social network sites: Definition, history, and scholarship. *Journal of Computer-Mediated Communication, 13*(1):article 11. http://jcmc.indiana.edu/vol13/issue1/boyd.ellison.html.

Brabazon, T. 2010. The Finns have got it right. *Times Higher Education Supplement,* August 11.

Bransford J., Brown A., and Cocking R. 2000. *How people learn: Brain, mind, experience and school.* Washington, DC: National Academy Press.

Bransford, J., Stevens, R., Schwartz, D., Meltzoff, A., Pea, R., Roschelle, J., Vye, N., Kuhl, P., Bell, P., Barron, B., Reeves, B., and Sabelli, N. 2006a. Learning theories and education: Toward a decade of synergy. In P. A. Alexander and P. H. Winne (eds.), *Handbook of educational psychology* (pp. 209–244). Mahwah, NJ: Lawrence Erlbaum.

Bransford, J., Barron, B., Pea, R., Meltzoff, A., Kuhl, P., Bell, P., Stevens, R., Schwartz, D., Vye, N., Reeves, B., Roschelle, J., and Sabelli, N. 2006b. Foundations and opportunities for an interdisciplinary science of learning. In K. Sawyer (ed.), *The Cambridge handbook of the learning sciences* (pp. 19–34). New York: Cambridge University Press.

Bromley, H. 1997. The social chicken and the technological egg: Educational computing and the technology/society divide. *Educational Theory*, *47*(1), 51–65.

Brown, A. 2009. Digital technology and education: Context, pedagogy and social relations. In R. Cowen and A. Kazamias (eds.), *International handbook of comparative education* (pp. 1159–1172). Berlin: Springer.

Brown, S., and Capdevila, R. 1999. Perpertuum mobile: Substance, force and the sociology of translation. In J. Law and J. Hassard (eds.), *Actor network theory and after* (pp. 26–50). Oxford, UK: Blackwell Publishers.

Buckingham, D. 2006. Is there a digital generation? In D. Buckingham and R. Willett (eds.), *Digital generations: Children, young people and new media*. Mahwah, NJ: Erlbaum.

Buckingham, D. 2007. *Beyond technology*. Cambridge, UK: Polity Press.

Burawoy, M. 2005. The critical turn to public sociology. *Critical Sociology*, 31, 3, pp.313–326.

Burn, A., Buckingham, D., Parry, R., and Powell, M. 2010. Minding the gaps, teachers' cultures, students' cultures. In D. Alvermann (ed.), *Adolescents' online literacies: Connecting classrooms, digital media, and popular culture* (pp. 183–201). New York: Peter Lang.

Bush, T. 2012. Three core elements to a successful BYOD strategy. [Accessed March 25, 2012] http://blogs.msdn.com/b/ukschools/archive/2012/02/24/3-core-elements -to-a-successful-byod-strategy.aspx.

Byron Review. 2008. *Safer children in a digital world: The report of the Byron Review*. Nottingham, UK: The Department for Children, Schools and Families, and the Department for Culture, Media and Sport.

Byron Review. 2010. *Do we have safer children in a digital world? A review of progress since the 2008 Byron Review*. Nottingham, UK: The Department for Children, Schools and Families Publications.

Calhoun, C. 2002. The class consciousness of frequent travelers: Toward a critique of actually existing cosmopolitanism. *The South Atlantic Quarterly*, *101*(4), 869–897.

Camicia, S. and Franklin, B. 2010. Curriculum reform in a globalised world: The discourses of cosmopolitanism and community. *London Review of Education*, 8(2), 93–104

Casella, R. 2010. Safety or social control? The security fortification of schools in a capitalist society. In T. Monahan and R. Torres (eds.), *Schools under surveillance: Cultures of control in public education* (pp. 73–86). New York: Rutgers University Press.

Castells, M. 1996. *The rise of the network society*. Oxford, UK: Blackwell.

Castells, M. 2009. *Communication power*. Oxford, UK: Oxford University Press.

Cavanagh, A. 2007. *Sociology in the age of the internet*. Buckingham: Open University Press.

Cervantes, R., Warschauer, M., Nardi, B., and Sambasivan, N. 2011. Infrastructures for low-cost laptop use in Mexican schools. In *Proceedings of the 29th International Conference on Human Factors in Computing Systems* (pp. 945–954). New York, ACM Conference on Human Factors in Computing Systems.

Chan, A. 2012. Hacking digital universalism: OLPC and information networks in the Andes. Paper presented to Second ISA Forum of Sociology, August. Buenos Aries.

Changizi, M. 2012. http://seedmagazine.com/content/article/humans_version_3.0/.

Chirillo, J. 2001. *Hack attacks encyclopedia*. London: John Wiley and Sons.

Chomsky, N. 1967. A review of B. F. Skinner's verbal behavior. In L. A. Jakobovits and M. S. Miron (eds.), *Readings in the psychology of language* (pp. 142–143). Upper Saddle River NJ: Prentice-Hall.

Chowdry, H., Crawford, C., and Goodman, A. 2009. Drivers and barriers to educational success: Evidence from the longitudinal study of young people in England. London: Institute for Fiscal Studies.

Christie, N. 1977. Conflicts as property. *British Journal of Criminology, 17*, 1–15.

Clandinin, J., and Huber, J. 2002. Narrative inquiry: Toward understanding life's artistry. *Curriculum Inquiry, 32*, 161–169.

Clarke, S. 1994. *Marx's Theory of crisis*. Basingstoke: Macmillan Press.

Collins, H., and Evans, R. 2002. The third wave of science studies: Studies of expertise and experience. *Social Studies of Science, 32*(2), 235–296.

Conle, C. 2000. Thesis as narrative or what is the inquiry in narrative inquiry. *Curriculum Inquiry, 30*, 189–214.

Conrad, J. 1963. *Nostromo: A tale of the seaboard*. London: Dent.

Common Sense Media (2010) *Internet Safety: Rules of the Road for Kids*. [Accessed May 10, 2013] www.commonsensemedia.org/advice-for-parents/rules-road-kids.

Coppens, P., Fernandes da Silva, M., and Pettman, S. 2006. European regulations on nutraceuticals, dietary supplements and functional foods. *Toxicology, 221*(1), 59–74.

Coveney, C., Gabe, J., and Williams, S. 2011. The sociology of cognitive enhancement: Medicalization and beyond. *Health Sociology Review, 20*(4), 381–393.

Cowie, R. 2005. What are people doing when they assign everyday emotion terms? *Psychological Inquiry, 16*(1), 11–48.

Cranmer, S., Selwyn, N., and Potter, J. 2009. Exploring primary pupils' experiences and understandings of "e-safety." *Education and Information Technologies, 14*(2), 127–142.

Cristiá, J., Ibarraran, P., Cueto, S., Santiago, A., and Severín, E. 2012. *Technology and child development: Evidence from the One Laptop per Child Program*. Washington, DC: IDB Working Paper Series 304.

Crook, C. 2008. Theories of formal and informal learning in the world of web 2.0. In S. Livingstone (ed.), *Theorising the benefits of new technology for youth*.Oxford: University of Oxford/London School of Economics, (pp. 30–37).

Cuban, L. 2001. *Oversold and underused: Computers in the classroom*. Cambridge, MA: Harvard University Press.

Cunill Grau, N. 2005. La intersectorialidad en el gobierno y gestión de la política social. Paper delivered *at X Congreso Internacional CLAD sobre la Reforma del Estado y de la Administración Pública*. Santiago de Chile. www.clad.org.ve.

Dale R., Robertson, S., and Shortis, T. 2004. You can't not go with the technological flow, can you? *Journal of Computer Assisted Learning, 20*, 456–470.

Davidson, C., and Goldberg, D. 2010. *The future of learning institutions in a digital age*. Cambridge, MA: MIT Press.

Davies, J. 2012. *Challenging governance theory: From networks to hegemony*. Bristol: Policy Press.

Davison, A. 2001. *Technology and the contested meanings of sustainability*. Albany, NY: State University of New York Press.

DCC. 2008. Children and Young People's Services press release. http://www. durham.gov.uk/durhamcc/pressrel.nsf/Web+Releases/9B151A656B3FD9AB802 574CF002D51F1?OpenDocument.

de Bastion, G., and Rolf, T. 2008. Low-cost ICT devices—new solutions for development? *Rural21*, June, 30–31.

de Lima, J. A. 2010. Thinking more deeply about networks in education. *Journal of Educational Change, 11*, 1–21.

De Saulles, M., and Horner, D. 2011. The portable panopticon: Morality and mobile technologies. *Journal of Information, Communication and Ethics in Society, 9*(3), 206–216.

Dean, M. 2010. *Governmentality: Power and rule in modern society* (2nd ed.). London: Sage.

Deleuze, G., and Guattari, F. 1984. *Anti-Oedipus: Capitalism and schizophrenia*. London: Athlone.

Department for Education and Employment [DfEE]. 1997. *The government's consultation paper. National grid for learning, connecting the learning society*. London: HMSO.

Department for Education and Skills [DfES]. 2002. *Transforming the way we learn: A vision for the future of ICT in schools*. London: DfES.

Department for Education and Skills [DfES]. 2003. *Fulfilling the potential: Transforming teaching and learning through ICT in schools*. London: DfES.

Departamento de Monitoreo y Evaluación del Plan Ceibal. 2011. *Encuesta Nacional a Docentes de Secundaria*. Montevideo: Plan Ceibal, Dirección Sectorial de Planificación Educativa.

DiMaggio, P., and Powell, W. 1983. The iron cage revisited: Institutional isomorphism and collective rationality in organizational fields. *American Sociological Review, 48*(2), 147–160.

Doll, W. 2008. Complexity and the culture of curriculum. *Educational Philosophy and Theory, 40*: 190–212.

Dotsub. 2007. *Transcript for One Laptop per Child*. http://dotsub.com/view /170314e6–34cc-4b64–84d2 483ac5bb800c/viewTranscript/eng.

Douglas, M. 1966. *Purity and danger*. London: Routledge and Kegan Paul.

Douglas, M., and Wildavsky, A. 1982. *Risk and culture: An essay on the selection of technological and environmental dangers*. Berkley, CA: University of California Press.

Dreyfus, H. 1995. Heidegger on gaining a free relation to technology. In A. Feenberg and A. Hannay (eds.), *Technology and the politics of knowledge* (pp. 97–107). Bloomington, IN: Indiana University Press.

Druin, A. 2002. The role of children in the design of new technologies. *Behaviour and Information Technology, 21*(1), 1–25.

228 • Bibliography

Duffy, T., and Cunningham, D. 1996. Constructivism, implications for the design and delivery of instruction. In D. H. Jonassen (ed.), *Handbook of research on educational communications and technology* (pp. 170–198). New York: Simon and Schuster Macmillan.

Durose, C. 2012. *Toward co-production in research with communities*. Swindon: Arts and Humanities Research Council, Connected Communities Scoping Study.

Dussel, I. 2011. *Enseñar y aprender con nuevas tecnologías*. Buenos Aires: Fundación Santillana.

Dussel, I., and Quevedo, L. 2010. *Educación y nuevas tecnologías: Los desafíos pedagógicos ante el mundo digital*. Buenos Aires: Fundación Santillana.

Dyer Witheford, N., and de Peuter, G. 2009. *Games of empire: Global capitalism and video games*. Minnesota, MN: University of Minnesota Press.

Eichstaedt, P. 2011. *Consuming the Congo: War and conflict minerals in the world's deadliest place*. Chicago: Lawrence Hill.

Erickson, M. 1991. *My voice will go with you*. New York: W. W. Norton and Co.

Erneling, C. 2010. *Toward discursive education: Philosophy, technology, and modern education*. Cambridge, UK: Cambridge University Press.

Ertmer, P., and Newby, T. 1993. Behaviorism, cognitivism, constructivism: Comparing critical features from an instructional design perspective. *Performance Improvement Quarterly*, 6(4), 50–70.

Eubanks, V. 2011. *Digital dead end: Fight for social justice in the information age*. Cambridge, MA: MIT Press.

Facer, K. 2011. *Learning futures*. London: Routledge.

Facer, K. 2012. After the moral panic? Reframing the debate about child safety online. *Discourse, Studies in the Cultural Politics of Education*, 33(3), 397–413.

Facer, K., and Green, H. 2007. Curriculum 2.0: Educating the digital generation. In S. Parker (ed.), *Unlocking innovation: Why citizens hold the key to public service reform*. London: Demos, (pp. 47–58).

Facer K., J. Furlong, R. Furlong and R. Sutherland, 2003. *ScreenPlay: Children and Computing in the Home*. London: RoutledgeFalmer.

Fairclough, N. 2000. Language and neo-liberalism. *Discourse and Society, 11*(2), pp. 147–148

Feenberg, A. 1999. *Questioning technology*. London: Routledge.

Feenberg, A. 2002. *Transforming technology: A critical theory revisited*. Oxford, UK: Oxford University Press.

Feenberg, A. 2011. *Between reason and experience: Essays in technology and modernity*. Cambridge, MA: MIT Press.

Fendler, L. 2001. Educating flexible souls: The construction of subjectivity through developmentality and interaction. In K. Hultqvist and G. Dahlberg (eds.), *Governing the child in the new millennium* (pp. 119–142). London: RoutledgeFalmer.

Fenwick, T., and Edwards, R. 2010. *Actor-network theory in education*. Abingdon: Routledge.

Ferguson, K., and Seddon, T. 2007. Decentred education: Suggestions for framing a socio-spatial research agenda. *Critical Studies in Education*, 48(1), 111–129.

Fielding, M., and Prieto, M. 2002. The central place of student voice in democratic renewal. In M. Schweisfurth, L. Davies, and C. Harber (eds.), *Learning democracy and citizenship: International experiences* (pp. 19–36). Oxford, UK: Symposium Books..

Fischer, F. 2000. *Citizens, experts, and the environment: The politics of local knowledge.* Durham, NC: Duke University.

Fisher, T. 2006. Educational transformation. *Education and Information Technologies, 11*(3–4), 293–303.

Fiske, J. 1990. *Introduction to communication studies.* New York: Routledge.

Foresight Mental Capital and Wellbeing Project. 2008. *Final project report— executive summary.* London: The Government Office for Science.

Foucault, M. 1985. *The history of sexuality.* New York: Vintage Books.

Foucault, M. 1988. *The care of the self: The history of sexuality,* volume 3. New York: Vintage Books.

Foucault, M. 1977. *Discipline and punish: The birth of the prison.* London: Allen Lane.

Foucault, M. 2008. *The birth of biopolitics.* London: Palgrave Macmillan.

Frankham, J. 2006. Network utopias and alternative entanglements for educational research and practice. *Journal of Education Policy, 21*(6), 661–677.

Freeden, M. 2003. *Ideology.* Oxford, UK: Oxford University Press.

Friesen, N. 2009. *Re-thinking e-Learning research: Foundations, methods and practices.* New York: Peter Lang.

Fuller, S. 2011. *Humanity 2.0: What it means to be past, present and future.* Basingstoke: Palgrave Macmillan.

Furedi, F. 2006. *Culture of fear revisited* (4th ed.). London: Continuum.

Gadamer, H. 2004. *Truth and method.* London: Continuum.

Gall, S., Beins, B., and Feldman, A. J. 1996. *The Gale encyclopedia of psychology.* Farmington Hills, MI: Gale Group.

Gee, J. 2007. *Video games and embodiment.* http://inkido.indiana.edu/aera%5F2007/.

Gee, J. 2008. *What video games have to teach us about learning and literacy* (2nd ed.). New York: Palgrave MacMillan.

Gelb, A., and Decker, C. 2012. Cash at your fingertips: Biometric technology for transfers in developing countries. *Review of Policy Research, 29*(1), 91–117.

Giddens, A. 2002. *Where now for new labour?* London: Polity Press.

Global Witness. 2009. *Faced with a gun, what can you do? War and the militarisation of mining in eastern Congo.* London: Global Witness.

Goldacre, B. 2009. *Bad science.* London: Harper Collins.

Goldmann, H. 2007. Reframing the debate. *Learning and Leading with Technology, 35*(4), 46.

Golumbia, D. 2009. *The cultural logic of computation.* Cambridge, MA: Harvard University Press.

Goodson, I. 1992. Computers in school as symbolic ideological action: The genealogy of the icon. *Curriculum Journal, 3*(3), 261–276.

Goodson, I. 2005. *Learning, curriculum and life politics.* Abingdon: Routledge.

Goodson, I., and Mangan, M. 1996. Computer literacy as ideology. *British Journal of Sociology of Education, 17*(1): 65–79.

Gough, N. 2002. Voicing curriculum visions. In W. Doll and N. Gough (eds.), *Curriculum visions*. New York: Peter Lang.

Greef, de Tjerk E., Arciszewski, H., and Neerincx, M. 2010. Adaptive automation based on an object-oriented task model: Implementation and evaluation in a realistic C2 environment. *Journal of Cognitive Engineering and Decision Making*, *4*(2), 152–182.

Greef, T., Kees van Dongen, K., Grootjen, M., and Lindenberg, J. 2007. Augmenting cognition: Reviewing the symbiotic relation between man and machine. In Schmorrow, D. and Reeves, L. (eds.), *Foundations of augmented cognition*. Berlin: Springer, (pp 439–448).

Gunter, H. 2009. The "C" word in educational research. *Critical Studies in Education*, *50*(1), 93–102.

Habermas, J. 1987. *Lifeworld and system: A critique of functionalist reason*. Boston, MA: Beacon Press.

Hall, I., and Higgins, S. 2005. Primary school students' perceptions of interactive whiteboards. *Journal of Computer Assisted Learning*, *21*(2), 102–117.

Hamm, S., and Smith, G. 2008. One laptop meets big business. *Business Week*, June 5.

Hammond, M., Younie, S., Woollard, J., Cartwright, V., and Benzie, D. 2009. *What does our past involvement with computers in education tell us?* Coventry, UK: Association for Information Technology in Teacher Education.

Hand, M., and Sandywell, B. 2002. E-topia as cosmopolis or citadel. On the democratizing and de-democratizing logics of the internet, or, toward a critique of the new technological fetishism. *Theory, Culture and Society*, *19*(1–2), 197–225.

Hardt, M., and Negri, A. 2000. *Empire*. Cambridge, MA: Harvard University Press.

Harris, J. 2010. *Enhancing evolution*. Princeton, NJ: Princeton University Press.

Harrison, C., Comber, C., Fisher, T., Haw, K., Lewin, C., Lunzer, E., McFarlane, A., Mavers, D., Scrimshaw, P., Somekh, B., and Watling, R. 2002. *ImpaCT2: The impact of information and communication technologies on pupil learning and attainment*. ICT in Schools Research and Evaluation Series. London: DfES.

Hartley, D. 1997. *Reschooling society*. London: RoutledgeFalmer.

Hartley, D. 2006. The instrumentalization of the expressive in education. In A. Moore (ed.), *Schooling, society and curriculum* (pp. 60–70). Abingdon: Routledge.

Harvey, D. 1990. *The condition of postmodernity: An enquiry into the origins of cultural change*. Cambridge: Blackwell.

Harvey, D. 2005. *A brief history of neoliberalism*. New York: Oxford University Press.

Harvey, D. 2010. *The enigma of capital*. London: Profile Books.

Harvey, D. 2002. Computers for the third world. *Scientific American*, *287*(4), 100–102.

Hasenbrink, U., Livingstone, S., Haddon, L., Kirwil, L., and Ponte, C. 2007. *EU kids go online: Comparing children's online activities and risks across Europe*. http://www.lse.ac.uk/collections/EUKidsOnline/Reports.

Haugsbakk, G., and Nordkvelle, Y. 2007. *The rhetoric of ICT and the new language of learning: A critical analysis of the use of ICT in the curricular field.* European Educational Research Journal, 6(1), 1–12.

HEFCE. 2012. *Enhancing learning and teaching through the use of technology.* [Accessed March 25, 2012] http://www.hefce.ac.uk/learning/techlearn/.

Heidegger, M. 1977. *The question concerning technology and other essays.* New York: Harper and Row, Publishers.

Held, D. 2010. *Cosmpolitansism: Ideals and realities.* Cambridge, UK: Polity.

Hergenhahn, B., and Olson, M. 2001. *An introduction to theories of learning* (6th ed.). Upper Saddle River, NJ: Prentice Hall.

Herrera, J. C. 2006. Monitoreo y Evaluación de Políticas Públicas, El Sistema de Información para la Gestión en Argentina. Paper presented at *XI CLAD Congress on State and Public Administration Reform*, Guatemala.

HESA. 2012. *Staff in higher education institutions.* Cheltenham: Higher Education Statistics Agency.

Higgins, S., Beauchamp, G., and Miller, D. 2007. Reviewing the literature on interactive whiteboards. *Learning, Media and Technology, 32*(3), 213–235.

Hlynka, D. & Belland, J. 1991. *Paradigms Regained. The uses of semiotic, postmodern and illuminative criticisms as modes of inquiry in educational technology.* Englewood Cliffs: Educational Technology Publications

HM Treasury. 2012. *Budget 2012.* [Accessed March 25, 2012] http://www.hm-treasury.gov.uk/budget2012.htm.

Hoadley, C. 2005. Design-based research methods and theory building: A case study of research with SpeakEasy. *Educational Technology, 45*(1), 42–47.

Hoffman, J. 2004. *Citizenship beyond the state.* London: Sage.

Holloway, J. 2002. *Change the world without taking power.* London: Pluto Press.

Holloway, J. 2010. *Crack capitalism.* London: Pluto Press.

Holloway, S. and Valentine, G. 2003 *Cyberkids: Children in the Information Age.* London: RoutledgeFalmer.

Hope, A. 2007. Risk-taking, boundary-performance and intentional school internet "misuse." *Discourse, Studies in the Cultural Politics of Education, 28*(1), 87–99.

Hope, A. 2011. *Internet risk @ school: Cultures of control in state education.* Saarbrücken: Lambert Press.

Høstaker, R. 2005. Latour: Semiotics and science studies. *Science Studies, 18*(2), 5–25.

House, E. 1999. *Values in evaluation and social research.* London: Sage.

iKeepSafe. 2011. *New e-rate requirements.* http://www.ikeepsafe.org/educators/e-rate/.

Ikonen, V., Kanerva, M., Kouri, P., Stahl, B. C., and Kutoma, W. 2010. *ETICA project deliverable D.1.2, emerging technologies report.* [Accessed March 25, 2012] http://ethics.ccsr.cse.dmu.ac.uk/etica/deliverables/D12Emergingtechnologies reportfinal.pdf.

IPCC [Intergovernmental Panel on Climate Change]. 2007. *Climate change 2007— the physical science basis.* Cambridge, UK: Cambridge University Press.

Ito, M. et al (eds.). 2010. *Hanging out, messing around, and geeking out: Kids living and learning with new media.* Cambridge, MA: MIT Press.

Jacobsen, M. 2004. Cognitive visualizations and the design of learning technologies. *International Journal of Learning Technology*, *1*(1), pp.40–62

James, J. 2011. Low-cost computers for education in developing countries. *Social Indicators Research*, *103*(3), 399–408.

Januszewski, A., and Molenda, M. 2007. *Educational technology: A definition with commentary* (2nd ed.). London: Routledge.

Jarvis, J. 2010. *iPad danger: App v. web, consumer v. creator*. [Accessed March 25, 2012] http://bit.ly/bKkuG6.

Jenkins, H. 2006. *Convergence Culture*. New York: NYU Press

Jenkins, H., Purushotma, R., Weigel, M., Clinton, K., and Robinson, A. 2009. *Confronting the challenges of participatory culture: Media education for the 21st century*. Cambridge, MA: MIT Press.

Jensen, C. B., and Lauritsen, P. 2005. Digital Denmark: IT reports as material-semiotic actors. *Science, Technology and Human Values*, *30*(3), 352–373.

JISC. 2012. *Mobile learning infokit*. [Accessed March 25, 2012] https://mobilelearning infokit.pbworks.com.

Johnson, N. 2009. Teenage technological experts' views of schooling. *Australian Educational Researcher*, *36*(1), 59–72.

Jonassen, D. 1994. Thinking technology. *Educational Technology*, 34 (3), 34–37.

Jonassen, D., Hannum, W., and Tessmer, M. 1989. *Handbook of task analysis procedures*. New York: Praeger.

Jonassen, D., Peck, K., and Wilson, B. 1999. *Learning with technology: A constructivist perspective*. Upper Saddle River, NJ: Merrill.

Jones, P. 2009. *World Bank financing of education: Lending, learning and development* (2nd ed.). London: Routledge.

Kable. 2008. *Education ICT in the UK public sector to 2012*. London: Kable.

Kelsoe, J. 2010. Behavioral neuroscience: A gene for impulsivity. *Nature*, *468*, 1049–1050.

Kenway, J. and Fahey, J. 2008. Imagining research otherwise. In Kenway, J. and Fahey, J. (eds.), *Globalizing the research imagination*. London: Routledge.

Kerr, S. 2003. Sociology and educational technology. In D. Jonnasen (ed.), *Handbook of research for educational communications and technology* (2nd ed.) (pp.72–120). Mahwah, NJ: Erlbaum.

Kirkpatrick, G. 2004. *Critical technology: A social theory of personal computing*. Aldershot: Ashgate.

Klebl, M. 2008. Explicating the shaping of educational technology, social construction of technology in the field of ICT and education. In *Readings in education and technology, proceedings of ICICTE 2008* (pp. 278–289). [Accessed May 10, 2013] www.icicte.org/ICICTE2008Proceedings/authorLIST.html.

Koehler, M., and Mishra, P. 2008. Introducing TPCK. In AACTE Committee on Innovation and Technology (eds.), *Handbook of technological pedagogical content knowledge (TPCK) for educators* (pp. 3–29). New York: Routledge.

Kozma, R. 1987. The implications of cognitive psychology for computer-based learning tools. *Educational Technology*, *27*(11):20–25

Kullman, K., and Lee, N. 2012. Liberation from/liberation within: Examining One Laptop per Child with Amartya Sen and Bruno Latour. In I. Oosterlaken and J. van den Hoven (eds.), *The capability approach: Technology and design philosophy of engineering and technology* (pp. 39–55). Berlin: Springer.

Kupchik, A., and Monahan, T. 2006. The New American School: Preparation for post-industrial discipline. *British Journal of Sociology of Education*, 27(5), 617–631.

Kvasny, L. 2006. Cultural (re)production of digital inequality in a US community technology initiative. *Information, Communication and Society*, 9(2), 160–181.

Laclau, E. 2005. *La razón populista*. Buenos Aires: FCE.

Lagos Céspedes, M., and Silva Quiróz, J. 2011. Estado de las experiencias 1 a 1 en Iberoamérica. *Revista Iberoamericana de Educación*, 56, 75–94.

Langemann, E. C. 1989. The plural worlds of educational research. *History of Education Quarterly*, 29(2), 185–214.

Lanier, J. 2010. *You are not a gadget*. New York: Vintage.

Lash, S. 2002. *Critique of information*. London: Sage.

Latour, B. 1987. Science in action: How to follow scientists and engineers through society. Cambridge, MA: Harvard University Press.

Latour, B. 1991. Technology is society made durable. In J. Law (ed.), *A sociology of monsters: Essays on power, technology and domination* (pp. 103–131). New York: Routledge.

Latour, B. 1993. *We have never been modern*. Cambridge, MA: Harvard University Press.

Lauder, H., Brown, P. and Halsey, A. 2009. Sociology of education: A critical history and prospects for the future. *Oxford review of education*, 35, 5, pp.569–585.

Laurillard, D. 2008. *Digital technologies and their role in achieving our ambitions for education*. London: Institute of Education.

Law, J. 1992. Notes on the theory of the actor-network: Ordering, strategy and heterogeneity. *Systems Practice*, 5(4), 379–393.

Law, J. 2008. On sociology and STS. *Sociological Review*, 56, 623–649.

Law, J. 2009. Actor Network Theory and material semiotics. In B. Turner (ed.), *The new Blackwell companion to social theory* (pp. 142–158). New York: Wiley-Blackwell.

Learning Futures. 2010. *Learning futures: Engaging schools*. London: Paul Hamlyn Foundation.

Lee, N., and Motzkau, J. 2012. The biosocial event: Responding to innovation in the life sciences. *Sociology* 46(3), 426–441.

Lee, N., and Motzkau, J. Forthcoming. Varieties of biosocial imagination: Reframing responses to climate change and antibiotic resistance? *Science, Technology and Human Values*. Pagination pending.

Leontev, A. 1978. *Activity, consciousness and personality*. Englewood Cliffs, NJ: Prentice Hall.

Lévinas, E. 1987. *Time and the other*. Pittsburgh, PA: Duquesne University Press.

Lévinas, E. 1998. *Totality and infinity*. Pittsburgh, PA: Duquesne University Press.

Levinson, M. 2010. *From fear to Facebook: One school's journey.* Washington, DC: International Society for Technology in Education.

Lightfoot, C. 1997. *The culture of adolescent risk-taking.* London: Guildford Press.

Livingstone, S. 2012. Critical reflections on the benefits of ICT in education. *Oxford Review of Education,* 38(1), 9–24.

López, A. (ed.). 2010. *Complementación productiva en la industria del software en los países del Mercosur, impulsando la integración regional para participar en el mercado global.* http://www.iadb.org/intal/intalcdi/PE/2009/03310.pdf.

Lovelock, J. 2000. *Gaia: A new look at life on earth.* Oxford, UK: Oxford Paperbacks.

Lucas, B., and Claxton, G. 2009. *Wider skills for learning: What are they, how can they be cultivated, how could they be measured and why are they important for innovation?* London: NESTA.

Luckin, R., Bligh, B., Manches, A., Ainsworth, S., Crook, C., and Noss, R. 2012. *Decoding learning: The proof, promise and potential of digital education.* London: NESTA.

Lupton, D. 1999. *Risk.* London: Routledge.

Luyt, B. 2008. The One Laptop per Child project and the negotiation of technological meaning. *First Monday,* 13, 6. [Accessed May 10, 2013] http://firstmonday.org/ojs/index.php/fm/article/view/2144/1971.

Lyotard, J. 1979. *The postmodern condition: A report on knowledge.* Manchester: Manchester University Press.

Mackinnon, R. 2011. *WikiLeaks, Amazon and the new threat to internet speech.* [Accessed March 25 2012] http://edition.cnn.com/2010/OPINION/12/02/mackinnon.wikileaks.amazon/.

Macris, V. 2011. The ideological conditions of social reproduction. *Journal for Critical Education Policy Studies,* 9(1) [Accessed March 3, 2012] www.jceps.com/PDFs/09-1-02.pdf.

Maggio, M., Lion, C., and Sarlé, P. 2012. *Creaciones, experiencias y horizontes inspiradores, la trama de Conectar Igualdad.* Ministerio de Educación de la Nación. http://bibliotecadigital.educ.ar/uploads/contents/CI_TIC_Trama0.pdf.

Mahatani, D., Debelle, R., Diallo, M. K., Dietrich, C. B., and Gramizzi, C. 2009. *Final report of the group of experts on the Democratic Republic of the Congo.* New York: UN Security Council document S/2009/603.

Mahiri, J. 2011. *Digital tools in urban schools; mediating a remix of learning.* Ann Arbor, University of Michighan Press.

Margolis, J. 2010. *Stuck in the Shallow End: Education, Race, and Computing.* Cambridge MA, MIT Press.

Margulis, L., and Sagan, D. 1997. *Microcosmos: Four billion years of microbial evolution.* Santa Cruz, CA: University of California Press.

Mansell, R. (2004) Political economy, power and new media. *New Media & Society,* 6, 1, (pp.96–105).

Markoff, J. 2005. *What the dormouse said: How the sixties counterculture shaped the personal computer.* London: Penguin.

Markoff, J. 2006. For $150, third-world laptop stirs big debate. *New York Times,* November 30. www.nytimes.com/2006/11/30/technology/30laptop.html.

Marx, K. 1993/1857. *Grundrisse: Foundations of the critique of political economy.* London: Penguin.

Marx, K. 2004/1867. *Capital, volume 1: A critique of political economy.* London: Penguin.

Marx, K. 2006/1885. *Capital, volume 2: A critique of political economy.* London: Penguin.

Masson, J. 1992. *Against therapy.* New York: Common Courage Press.

Matthewman, S. 2011. *Technology and social theory.* London: Sage.

McCarthy, H., Miller, P., and Skidmore, P. (eds.). 2004. *Network logic: Who governs in an interconnected world?* London: Demos.

McLennan, G. 2004. Travelling with vehicular ideas: The case of the third way. *Economy and Society, 33*(4): 484–499.

McWilliam, E. 1998. Created thinking in teacher education: The appeal of non-stupid optimism. *ATEA Conference Proceedings, Australian Teacher Education Association Conference.* July. Melbourne.

Meighan, R., and Siraj-Blatchford, I. 2003. *A sociology of educating* (4th ed.). London: Continuum.

Meiksins-Wood, E. 1997. Back to Marx. *Monthly Review, 49*(2), 1–9.

Millea, J., Galatis, H., and McAllister, A. 2009. *Annual report on emerging technologies: Planning for change,* SICTAS report, Education.au Limited, Adelaide, Australia. [Accessed March 09, 2012] http://www.educationau.edu.au/jahia/Jahia/home/SICTAS/pid/852.

Miller, D. 2011. *Tales from Facebook.* Cambridge: Polity.

Miller, P., and Rose, N. 2008. *Governing the present: Administering economic, social and personal life.* Cambridge, UK: Polity.

Mills, C. 1959/2000. *The sociological imagination.* Oxford, UK: Oxford University Press.

Ministerio de Educación de la Nación Argentina. 2009. *Lineamientos Políticos y Estratégicos de la Educación Secundaria Obligatoria.* Resolución Consejo Federal de Educación No. 84/09. http://portal.educacion.gov.ar/nueva-escuela-secundaria/.

Ministerio de Educación de la Nación Argentina. 2011a. *Estrategia político pedagógica y marco normativo del programa Conectar Igualdad.* http://bibliotecadigital.educ.ar/uploads/contents/estrategia0.pdf.

Ministerio de Educación de la Nación Argentina. 2011b. *Nuevas voces, nuevos escenarios, estudios evaluativos sobre el Programa Conectar Igualdad.* Buenos Aires: DINIECE. www.me.gov.ar.

Ministerio de Educación de la Nación Argentina. 2012. *Conectar Igualdad en las Escuelas. Línea de Base 2011.* Internal Draft. Buenos Aires.

Monahan, T. 2005. *Globalization, technological change, and public education.* Abingdon: Routledge.

Mosco, V. 2009. *The Political Economy of Communication.* London: Sage.

Moss, G., Jewitt, C., Levaãiç, R., Armstrong, V., Cardini. A., and Castle, F. 2007. *The interactive whiteboards: Pedagogy and pupil performance evaluation.* London: Institute of Education, University of London/DfES.

Mukul, A. 2006. HRD rubbishes MIT's laptop scheme for kids. *Times of India*, July 3.

Mulgan, G. 2006. Thinking in tanks: The changing ecology of political ideas. *The Political Quarterly*, 77(2): 147–155.

NAACE. 2008. *ICT mark*. [Accessed March 3, 2012] http://www.naace.co.uk/ictmark.

National College of School Leadership [NCSL]. 2004. *Self evaluation: Models, tools and examples of practice.* Nottingham: NCSL.

Naughton, J. 2005. The $100 laptop question. *The Observer*, Business supplement December 4, p. 6.

Negri, A. 1989. *The politics of subversion: A manifesto for the twenty first century.* Cambridge, UK: Polity Press.

Negroponte, N. 2007. Keynote address to Internet and Society conference— "University—Knowledge beyond Authority." May 31 to June 1. Cambridge, MA: Berkman Centre for Internet and Society at Harvard Law School.

Negroponte, N. 2009. "Lessons learned and future challenges" speech given to *Reinventing the classroom: social and educational impact of information and communications technologies in education* seminar. September. Washington: Inter-American Development Bank.

Newman, F., and Holzman, L. 1997. *End of knowing: New developmental ways of learning.* London: Routledge.

Noble, D. 1984. Computer literacy as ideology. *Teacher College Record, 85*(4), 602–615.

Noble, D. 1998. Digital diploma mills: The automation of higher education. *First Monday.* [Accessed March 3, 2012] http://communication.ucsd.edu/dl/ddm1.html.

Nolin, J. 2010. *Speedism, boxism, and markism: Three ideologies of the internet.* [Accessed March 25, 2012] http://firstmonday.org/htbin/cgiwrap/bin/ojs/index.php/fm/article/viewArticle/2566/2630.

Northwest Center for Philosophy for Children. 2008. *Plato's allegory of the cave and* The matrix. [Accessed May 24, 2012] http://depts.washington.edu/nwcenter/lessonsplansplatoallegory.html.

Nutt, J. 2010. *Professional educators and the evolving role of ICT in schools: Perspective report.* CfBT Educational Trust. [Accessed March 15, 2012] www.ictliteracy.info/rf.pdf/ICTinSchools.pdf.

Nye, D. 1996. *American technological sublime.* Boston, MA: MIT Press.

Nye, D. 2007. *Technology matters: Questions to live with.* Cambridge, MA: MIT Press.

O'Neill, B., and Hagen, I. 2009. Media literacy. In S. Livingstone and L. Haddon (eds.), *Kids online: Opportunities and risks for children* (pp. 229–239). Bristol: The Policy Press.

Oakley, A. 2000. *Experiments in knowing.* Cambridge, UK: Polity.

Ofsted [The UK Office for Standards in Education, Children's Services and Skills]. 2010. *The safe use of new technologies.* http://www.ofsted.gov.uk/Ofsted-home/Publications-and-research/Browse-all-by/Documents-by-type/Thematic-reports/The-safe-use-of-new-technologies.

Oliver, M. 2011. Technological determinism in educational technology research: Some alternative ways of thinking about the relationship between learning and technology. *Journal of Computer Assisted Learning*, 27(5), 373–384.

OLPC [One Laptop per Child]. 2010. *One Laptop per Child—mission statement.* www.laptop.org/ vision.

On Guard Online. 2009. *Net cetera: Chatting with kids about being online.* Washington DC: US Department of Commerce.

Osborne, T. 2004. On mediators: Intellectuals and the ideas trade in the knowledge society. *Economy and Society*, 33(4), 430–447.

Ostrom, E. 1996. Crossing the great divide: Co-production, synergy and development. *World Development*, 24(6), 1073–1088.

Oxford English Dictionary [OED]. n.d. Learning. http://dictionary.oed.com.

Ozga, J. 2009. Governing education through data in England: From regulation to self-evaluation. *Journal of Education Policy*, 24(2), 149–163.

Padmanabhan, P., and Wise, A. 2012. Exploring situational factors shaping access in a laptop program for socially disadvantaged children in India: A case study. *Educational Media International*, 49(2), 81–95.

Pal, J., Patra, R., Nedevschi, S., Plauche, M., and Pawar, U. 2009. The case of the occasionally cheap computer, low-cost devices and classrooms in the developing world. *Information Technologies and International Development*, 5(1), 49–64.

Papacharissi, Z. 2010. *A networked self: Identity, community, and culture on social network sites.* New York: Routledge.

Papacharissi, Z. 2011. *A networked self-identity, community, and culture on social network sites.* London: Routledge.

Papert, S. 1980. *Mindstorms, children, computers, and powerful ideas.* New York: Basic Books.

Partnership for 21st Century Skills. 2009. *Curriculum and instruction: A 21st Century Skills implementation guide.* Tucson, AZ: Partnership for 21st Century Skills. http://www.21stcenturyskills.org/documents/p21-stateimp_curriculuminstruction.pdf.

Pashler, H., McDaniel, M., Rohrer, D., and Bjork, R. 2008. Learning styles: Concepts and evidence. *Psychological Science in the Public Interest*, 9(3), 105–119.

Patton, A. 2012. *Work that matters: The teacher's guide to project-based learning.* London: Paul Hamlyn Foundation.

Pérez Burger, M. 2009. *Evaluación Educativa del Plan Ceibal 2009.* Montevideo: Plan Ceibal, Dirección Sectorial de Planificación Educativa.

Pflaum, W. 2004. *The technology fix.* Alexandria, VA: Association for Supervision and Curriculum Development.

Philip, K., Irani, L., and Dourish, P. 2012. Postcolonial computing: A tactical survey. *Science, Technology, and Human Values*, 37(1), 3–29.

Piaget, J. 1927. *The child's conception of the world.* London: Routledge and Kegan Paul.

Pinar, W. 2004. *International handbook of curriculum research.* Mahwah, NJ: Lawrence Erlbaum.

Popkewitz, T. 2008. *Cosmopolitanism and the age of school reform: Science, education, and making society by making the child*. Abingdon: Routledge.

Popkewitz, T. 2012. Numbers in grids of intelligibility: Making sense of how educational truth is told. In H. Lauder, M. Young, H. Daniels, M. Balarin, and J. Lowe (eds.), *Educating for the knowledge economy? Critical perspectives* (pp. 169–191). Abingdon: Routledge.

Popkewitz, T., and Bloch, M. 2001. Administering freedom: A history of the present. In K. Hultqvist and G. Dahlberg (eds.), *Governing the child in the new millennium* (pp. 85–118). London: RoutledgeFalmer.

Popkewitz, T., Olsson, U., and Petersson, K. 2006. The learning society, the unfinished cosmopolitan, and governing education. *Educational Philosophy and Theory*, 38(4), 431–449.

Prensky, M. 2008. Turning on the lights. *Educational Leadership*, 65(6), 40–45.

Prensky, M. 2010. *Teaching digital natives*. London: Corwin/Sage.

Price, D. 2010. *Learning futures: Engaging students*. London: Paul Hamlyn Foundation/Innovation Unit.

Price, D. 2011. Learning futures: Rebuilding curriculum and pedagogy around student engagement. *Forum*, 53(2), 273–284.

Pritchard, A. 2008. *Ways of learning*. London: Routledge.

Raina, P., Austen, I., and Timmins, H. 2012. An idea promised the sky, but India is still waiting. *New York Times*, December 29.

Ready, R., and Burton, K. 2010. *Neuro-linguistic programming for dummies*. London: John Wiley and Sons.

Rivas, A., Vera, A., and Bazem, P. 2010. *Radiografía de la educación Argentina*. Buenos Aires: CIPPEC-Fundación Noble.

Rivoir, A. 2010. *El Plan Ceibal: Impacto comunitario e inclusión social, 2009–2010*. Uruguay: Facultad de Ciencias Sociales, Universidad de la República.

Rizvi, F. 2009. Toward cosmopolitan learning. *Discourse, Studies in the Cultural Politics of Education*, 30(3), 253–268.

Rizvi, F., and Lingard, B. 2010. *Globalizing education policy*. Abingdon: Routledge.

Robinson, L. 2009. A taste for the necessary: A Bourdieuian approach to digital inequlaity. *Information, Communication and Society*, 12(4), 488–507.

Rose, N. 1989. *Governing the soul*. London: Routledge.

Rose, N. 1999a. *Governing the soul: The shaping of the private self* (2nd ed.). London: Free Association Books.

Rose, N. 1999b. *Powers of freedom: Reframing political thought*. Cambridge, UK: Cambridge University Press.

Rose, N. 2007. *The Politics of life itself: Biomedicine, power and subjectivity in the twenty-first century*. Princeton, NJ: Princeton University Press.

Rowell, L. 2007. Can the $100 laptop change the world? *eLearn Magazine*. http://elearnmag.acm.org.

Rudd, T. 2007 *Interactive whiteboards in the classroom*. Bristol: Futurelab.

Ryan, J. 2010. *A history of the Internet and the digital future*. London: Reaktion.

Saettler, P. 2004. *The evolution of American educational technology* (2nd ed.). Mahwah, NJ: Lawrence Erlbaum.

Sagol, C. 2011. *El modelo 1 a 1. Notas para comenzar. Serie Estrategias en el aula para 1 a 1.* Buenos Aires: Ministerio de Educación.

Salen, K., Torres, R., Wolozin, L., Rufo-Tepper, R., and Shapiro, A. 2011. *Quest to learn: Developing the school for digital kids.* Cambridge, MA: MIT Press.

Sandel, M. 2006. *The case against perfection.* Harvard, MA: Harvard University Press.

Savvas, A. 2011. Forrester: School tech spending a "bright spot" in 2011. *Computerworld.* [Accessed March 3, 2012] www.computerworlduk.com/news/public-sector/3259425/forrester-school-tech-spending-a-bright-spot-in-2011/##.

Sawyer, R. 2006. Conclusion: The schools of the future. In R. K. Sawyer (ed.) *The Cambridge handbook of the learning sciences* (pp. 567–580). Cambridge, UK: Cambridge University Press.

Sawyer, T. 2000. *The Cambridge handbook of the learning sciences.* Cambridge: Cambridge University Press.

Scardamalia, M. 2003. Knowledge forum (Advances beyond CSILE). *Journal of Distance Education, 17*(3), 23–28.

Schank, Roger. 2000. Afterword: After how comes what. In Sawyer, T. (ed.) *'The Cambridge handbook of the learning sciences.* Cambridge: Cambridge University Press.

Schank, R. 2011. *Teaching minds: How cognitive science can save our schools.* New York: Teachers College Press.

Schofield, J. 1995. *Computers and classroom culture.* Cambridge, UK: Cambridge University Press.

Scott, D. 2008. *Critical essays on major curriculum theorists.* London: Routledge.

Selwyn, N. 1997. The continuing weakness of educational computing research. *British Journal of Educational Technology, 28*(4), 305–307.

Selwyn, N. 2007. The use of computer technology in university teaching and learning. *Journal of Computer Assisted Learning, 23*(2), 83–94.

Selwyn, N. 2008a. Realising the potential of new technology? Assessing the legacy of New Labour's ICT agenda 1997–2007. *Oxford Review of Education, 34*(6), pp.701–712

Selwyn, N. 2008b. Developing the technological imagination. In S. Livingstone (ed.), *Theorising the benefits of new technology for youth.* University of Oxford/LSE, pp. 18–19.

Selwyn, N. 2010. Looking beyond learning: Notes toward the critical study of educational technology. *Journal of Computer Assisted Learning, 26,* 65–72.

Selwyn, N. 2011a. *Education and technology: Key issues and debates.* London: Continuum.

Selwyn, N. 2011b. *Schools and schooling in the digital age: A critical analysis.* Abingdon: Routledge.

Selwyn, N. 2011c. The place of technology in the Conservative-Liberal Democrat education agenda: An ambition of absence? *Educational Review, 63*(4). pp.395–408.

Selwyn, N. 2012. Making sense of young people, education and digital technology: The role of sociological theory. *Oxford Review of Education, 38*(1), 81–96.

Selwyn, N., Potter, J., and Cranmer, S. 2010. *Primary schools and ICT: Learning from pupils perspectives*. London: Continuum.

Shirky, C. 2008. *Here comes everybody*. London: Allen Lane.

Shutkin, D. 1998. The deployment of information technology and the augmentation of the child. In T. Popkewitz and M. Brennan (eds.), *Foucault's challenge, discourse, knowledge and power in education* (pp. 205–229). New York: Teachers College Press.

Shutkin, D. 2005. Neoliberalism, the technological sublime, and techniques of the self. *Educational Technology*, *45*(2), 39–48.

Sileoni, A. 2012. Inaugural words. In Dussel, I. (ed.), *Tic y Educación, Aprender y enseñar en la cultura digital* (pp. 73–77). Buenos Aires: Fundación Santillana.

Simon, H. 1969. *The sciences of the artificial*. Cambridge, MA: MIT Press.

Simpson, L. 1995. *Technology, time and the conversations of modernity*. London: Routledge.

Skinner, B. 1968. *The technology of teaching*. New York: Appleton-Century-Crofts.

Smith, M. 1994. Recourse of empire. In M. Smith and L. Marx (eds.), *Does technology drive history? The dilemma of technological determinism*. Cambridge, MA: MIT Press. (pp.36–53)

Smits, M. 2001. Langdon winner: Technology as a shadow constitution. In Achterhuis, H. (ed.), *American philosophy of technology: The empirical turn*. Bloomington IN, Indiana University Press (pp.147–169).

Somekh, B. 2007. *Final report of the evaluation of the ICT test bed project*. Coventry, UK: Becta.

Staksrud, E. 2009. Problematic conduct: Juvenile delinquency on the internet. In S. Livingstone and L. Haddon (eds.), *Kids online: Opportunities and risks for children* (pp. 147–157). Bristol: The Policy Press.

Stevenson, H. 2011. Coalition education policy: Thatcherism's long shadow. *FORUM*, *53*(2) [Accessed February 1, 2012] www.wwwords.co.uk/FORUM.

Stiegler, B. 2009. *For a New Critique of Political Economy*. Cambridge: Polity.

Strathern, M. 1996. Cutting the network. *Journal of the Royal Anthropological Institute*, *2*, 517–535.

Streicher-Porte, M., Marthaler, C., Boni, H., Schuep, M., Camacho, A., and Hilty, L. 2009. One laptop per child, local refurbishment or overseas donations? Sustainability assessment of computer supply scenarios for schools in Columbia. *Journal of Environmental Management*, *90*(11), 3498–3511.

Strhan, A. 2007. Bringing me more than I contain: Discourse, subjectivity and the scene of teaching in Totality and infinity. *Journal of Philosophy of Education*, *41*, 411–430.

Stringer, C. 2011. *The origin of our species*. London: Allen Lane.

Strom, S. 2010. Non-profits review technology failures. *New York Times*, August 16.

Suoranta, J., and Vadén, T. 2010. *Wikiworld*. London: Pluto Press.

Sutherland, R. 2004. Designs for learning: ICT and knowledge in the classroom. *Computers & Education*, *43*, 5–16.

Tabb, L. 2008. A chicken in every pot—One Laptop per Child, the trouble with global campaign promises. *E-Learning and Digital Media*, *5*(3), 337–351.

Tao, J., and Tan., T. 2005. Affective computing: A review. *Lecture Notes in Computer Science, 3784*, 981–995.

Thaler, R. H., and Sunstein, C. R. 2009. *Nudge: Improving decisions about health, wealth and happiness.* London: Penguin.

Thomas, J. 2003. Facilitation of critical thinking and deep cognitive processing by structured discussion board activities. *Teaching in Higher Education Forum E-Proceedings. Centers for Excellence in Learning and Teaching.* Louisiana State University. http://www.celt.lsu.edu/CFD/E-Proceedings/Facilitation%20of%20 Critical%20Thinking%20and%20Deep%20Cognitive%20Processing.htm.

Thompson, J. 1995. *The media and modernity.* Palo Alto, CA: Stanford University Press.

Thomson, P., Jones, K., and Hall, C. 2009. *Creative school change.* Newcastle, UK: Creativity, Culture and Education.

Thorndike, E. 1901. *The Human nature club: An introduction to the study of mental life.* New York: Longmans Green and Co.

Thorndike, E. 1910. The contribution of psychology to education. *The Journal of Educational Psychology,* (1), 5–12.

Thorndike, E. 1911. *Animal intelligence.* New York: Macmillan.

Thorndike, E. 1912. *Education: A first book.* New York: MacMillan.

Tiqqun. 2001. *The cybernetic hypothesis.* [Accessed March 25, 2012] http://www. archive.org/details/Tiqqun1.

Triodos Bank. 2011. *Company engagement report.* [Accessed March 25, 2012] http:// www.triodos.com/downloads/research/company-engagement-report-2011.pdf.

Trucano, M. 2005. *Knowledge maps: ICT in education.* Washington, DC: infoDev/ World Bank.

Turmel, A. 2008. *A historical sociology of childhood.* Cambridge, UK: Cambridge University Press.

Uden, L., and Beaumont, C. 2006. *Technology and problem-based learning.* Hershey, PA: Information Science Publishing.

UKCCIS [UK Council for Child Internet Safety]. 2009. *Click clever click safe: The first UK child Internet safety strategy.* Nottingham: DCSF Publications.

UNESCO. 2007. *ICT-in-Education Toolkit for education policy makers, planners and practitioners.* [Accessed March 25, 2012] http://www.ictinedtoolkit.org/usere/ conceptblueprint.php.

UNESCO-IIPE-PNUD. 2009. *Abandono escolar y políticas de inclusión en la edu-cación secundaria.* http://www.oei.es/pdf2/abandono_escolar_politicas_inclu-sion.pdf.

US Department of Education. 2010. *National educational technology plan.* Washington DC: US Department of Education.

US Internet Safety Technical Task Force. 2008. *Enhancing child safety and online technologies.* Washington DC: US Internet Safety Technical Task Force.

Vacchieri, A., and Castagnino, L. 2012. *Historias uno a uno. Imágenes y testimonios de Conectar Igualdad.* Buenos Aires: Programa Conectar Igualdad.

Van Loon, J. 2002. *Risk and technological culture: Toward a sociology of virulence.* New York: Routledge.

Velkey, R. 2002. *Being after Rousseau: Philosophy and culture in question*. Chicago: University of Chicago Press.

Villanueva-Mansilla, E., and Olivera, P. 2012. Institutional barriers to development innovation: Assessing the implementation of XO-1 computers in two peri-urban schools in Peru. *Information Technologies and International Development*, 8(4), 177–189.

Viñao Frago, A. 2002. *Sistemas educativos, culturas escolares y reformas*. Madrid: Morata Editorial.

Virilio, P. 2003. *Unknown Quantity*. New York: Thames & Hudson.

Virno, P. 2001. *Generation online*. [Accessed March 25, 2012] http://www.generation -online.org/p/fpvirno10.htm.

Virno, P. 2004. *A grammar of the multitude: For an analysis of contemporary forms of life*. Los Angeles, CA: Semiotext.

Vygotsky, L. S. 1978. *Mind in society: The development of higher psychological processes*. Cambridge, MA: Harvard University Press.

Waller, T. 2007. ICT and social justice. *Journal for Critical Education Policy Studies*, 5(1). [Accessed 05.10.2013] www.jceps.com/?pageID=article&articleID=92.

Ware, N. (2001) *Congo war and the role of Coltan*. [Accessed May 10, 2013] www1. american.edu/ted/ice/congo-coltan.htm.

Warschauer, M., and Ames, M. 2010. Can one laptop per child save the world's poor? *Journal of International Affairs*, 64(1), 33–51.

Warschauer, M., Cotton, S., and Ames, M. 2011. One Laptop per Child Birmingham: Case study of a radical experiment. *International Journal of Learning and Media*, 3(2), 61–76.

Watson, D. (ed.). 1993. *The impact report*. London: King's College, Centre for Educational Studies.

Watson, D. 2001. Pedagogy before technology. *Education and Information Technologies*, 6(4), 251–266.

Webster, F. 2005. Making sense of the information age. *Information, Communication and Society*, 8(4): 439–458.

Weyland, K., Madrid, R., and Hunter, W. 2010. *Leftists governments in Latin America: Successes and shortcomings*. Cambridge, UK: Cambridge University Press.

Winner, L. 1986. *The Whale and the Reactor*. Chicago IL: University of Chicago Press.

Whitehead, A. 1927/1985. *Process and reality*. New York: The Free Press.

Williams, R. 1958. *Culture and society*. Harmondsworth: Penguin.

Williams, R. 1974. *Television, technology and cultural form*. London: Fontana.

Williamson, B. 2012. Centrifugal schooling: Third sector policy networks and the reassembling of curriculum policy in England. *Journal of Education Policy*, 27, 6, pp.775–794.

Wolinsky, A. 2008. We can get there from here: Realizing educational technology's potential in the face of internet safety issues. *MultiMedia and Internet @ Schools*, 15(4), 26–30.

Woolgar, S. 2002. *Virtual Society*. Oxford: Oxford University Press.

Yapp, C. 2002. "The learning renaissance" presentation to the Learning Lab Conference. June 26. Telford.

Young, M. 1984. Information technology and the sociology of education: Some preliminary thoughts. *British Journal of the Sociology of Education*, 5(2), 205–210.

Young, M. 1998. *The curriculum of the future: From the "new sociology of education" to a critical theory of learning*. London: Falmer Press.

Young, M. 2008. *Bringing knowledge back in: From social constructivism to social realism in the sociology of education*. Abingdon: Routledge.

Young, M., DePalma, A., and Garrett, S. 2002. Situations, interaction, process and affordances: An ecological psychology perspective. *Instructional Science, 30*, 47–63.

Yowell, C. 2012. Connected Learning: Designed to mine the new, social, digital domain. *DMLcentral.net*. March 1, 2012. [Accessed March 6, 2012] http://www.dmlcentral.net/blog/constance-m-yowell-phd/connected-learning-designed-mine-new-social-digital-domain.

Yujuico, E., and Gelb, B. 2011. Marketing technological innovation to LDCs, lessons from one laptop per child. *California Management Review, 53*(2), 50–68.

Zhao, Y., and Frank, K. 2003. Factors affecting technology uses in schools: An ecological perspective. *American Education Research Journal, 40*(1), 807–840.

Zhou, F., Been-Lirn, H., and Billinghurst, M. 2008. Trends in augmented reality tracking, interaction and display: A review of ten years of ISMAR. *ISMAR "08" Proceedings of the 7th IEEE/ACM International Symposium on Mixed and Augmented Reality*.

Žižek, S. 2008. *In defense of lost causes*, London: Verso.

Žižek, S. 2009. *First as tragedy, then as farce*. London: Verso.

Zournazi, M. 2002. *Hope: New Philosophies for Change*. London: Pluto.

Contributors

Ines Dussel (Mexico) is a principal investigator at the Departamento de Investigaciones Educativas, CINVESTAV (Centro de Investigación y de Estudios Avanzados del Instituto Politécnico Nacional). Ines holds a PhD from the University of Wisconsin-Madison. She works on the effects of digital media and visual culture in schooling, particularly on curriculum and social interaction in classrooms. Recent publications include *Aprender y enseñar en la cultura digital* (2011, Santillana) and *Escuelas, tecnologías y cultura visual* (2013, UNIPE).

Keri Facer (UK) is professor of educational and social futures at the University of Bristol, Graduate School of Education. She works on rethinking the relationship between formal educational institutions and wider society, particularly in relation to environmental, social, and technological disruption. Recent publications include *Learning Futures: Education, Technology and Social Change* (2011, Routledge).

Patricia Ferrante (Argentina) is a doctoral student working at Facultad Latinoamericana de Ciencias Sociales (FLACSO), Latin American School of Social Sciences, UNESCO. Patricia holds a degree in political science from the University of Buenos Aires, and has an MA in international relations at FLACSO-University of San Andrés.

Norm Friesen (Canada) is Canada Research Chair in E-Learning Practices, Thompson Rivers University. Norm's research interests include media theory, alternative pedagogies, technical e-learning standardization, phenomenology, and ethnomethodology. Recent publications include *Re-Thinking E-Learning Research: Foundations, Methods and Practices* (2009, Peter Lang) and *The Place of the Classroom and the Space of the Screen* (2011, Peter Lang). Norm has also recently edited and translated a German classic on modern education, its forms and cultures, *Forgotten Connections: On Culture and Upbringing* by Klaus Mollenhauer (2013, Peter Lang).

Richard Hall (UK) is De Montfort University's head of Enhancing Learning through Technology. He is a UK National Teaching Fellow and a Professor in Education and Technology. Richard is also a research associate in the Centre for Computing and Social Responsibility at De Montfort University. His research interests include the idea of the university and radical alternatives to it; technology and critical social theory; resilient education and the place of cooperative practice in overcoming disruption in higher education; and the place of social media in the idea of the twenty-first-century university. Richard writes at http://richard-hall.org.

Andrew Hope (Australia) is an associate professor at the School of Education, University of Adelaide. His current research interests include critical explorations of education technology; schools and social control; young people, risk, and surveillance; representation, authority, and power in qualitative social research. Recent publications include *Internet Risk @ School: Cultures of Control in State Education* (2011, Lambert Press).

Nick Lee (UK) is an associate professor at Institute of Education, University of Warwick. His core concerns are with the "separability" effects that distinguish "persons" within social, biological, and technological contexts and in the constitutive ambiguities of biosocial and biopolitical experience. Among publications that address these issues is a series of books focused on childhood considered as a global biopolitical phenomenon are *Childhood and Society: Growing up in an Age of Uncertainty* (2001, Open University Press), *Childhood and Human Values: Development, Separation and Separability* (2005, Open University Press), and *Childhood and Bio-Politics: Life Processes, Climate Change and Human Futures*, which he is currently writing for Palgrave Macmillan. The recent Economic and Social Research Council (ESRC)-funded project "Mimetic Factors in Individual Behaviour: Health and Well-Being" brought a network of sociologists, medics, philosophers, life scientists, and engineers together to examine the basic sociological issue of the susceptibility/ability of persons to influence each other.

Johanna Motzkau (UK) is a lecturer in psychology at the Faculty of Social Sciences, The Open University. She has a background in philosophy, German kritische psychologie, theoretical psychology, developmental psychology, and forensic psychology. Drawing on the work of process thinkers such as Bergson, Deleuze, Stengers, and Whitehead, her work explores the methodological value of process theory for conceptualizing the relationship between theory, research, and practice. She is interested in issues of memory and suggestibility, children's rights, child sexual abuse, gender, and the way in which psychological knowledge is used by the law. Recent research has looked at the history and theory of suggestibility research, and compared child witness practice in England and Germany. Publications include *Exploring the Transdisciplinary Trajectory of Suggestibility* (2009, Subjectivity) and *Around the Day in Eighty Worlds: Deleuze, Suggestibility and Researching Practice as Process* (2011, Captus Press).

Timothy Rudd (UK) is principal lecturer at the School of Education, University of Brighton. Tim's current research interests include the politics and ideology of educational technology; the use of technology for social good; innovation and transformation in education; and alternative educational approaches and possibilities. Tim is currently working on a number of projects relating to education and the use of new technologies in both formal and informal settings.

Julian Sefton-Green (UK) is currently principal research fellow at the Department of Media & Communication, London School of Economics and Politics (LSE) and a research associate at the University of Oslo. He is an Honorary Professor of Education at the University of Nottingham, UK, and at the Hong Kong Institute of Education. Julian has worked as a schoolteacher, in teacher training and in the informal education sector. He has researched and written widely on many aspects of media education, creativity, new technologies, and informal learning. Recent volumes include joint editing of *The International Handbook of Creative Learning* (2011, Routledge) and *Learning Lives: Transactions, Digital culture and Learner Identity* (2013, Cambridge University Press). Julian writes at www.julianseftongreen.net.

Neil Selwyn (Australia) works at the Faculty of Education, Monash University, where his research and teaching focuses on education, technology, and society. Recent publications include *Education in a Digital World: Global Perspectives on Technology and Education* (2013, Routledge), *Schools and Schooling in the Digital Age: A Critical Perspective* (2011, Routledge), and *Education and Technology: Key Issues and Debates* (2011, Bloomsbury).

David Shutkin (USA) is an associate professor of educational technology and a member of the faculty in the Department of Education and Allied Studies at John Carroll University. David's scholarship concerns technology studies and nonfoundational ethics in education. At present, he is researching the lived experiences of children and technology in schools. Methodologically, David employs forms of discourse analysis, actor network theory, and postphenomenology.

Ben Williamson (UK) is a lecturer at the School of Education, University of Stirling. Ben's research focuses mainly on curriculum change and the educational uses of digital media. He is especially concerned with following new policy actors and curriculum developers such as those from cross-sectoral organizations, new "policy networks," and the new "edu-experts" of think tanks, nonprofits, consultancies, semigovernmental spin-offs, and commercial philanthropies. Recent publications include *The Future of the Curriculum: School Knowledge in the Digital Age* (2013, MIT Press) and *Learning Identities in a Digital Age: Rethinking Creativity, Technology and Education* (with Avril Loveless—2013, Routledge).

Index

GPSR Compliance

The European Union's (EU) General Product Safety Regulation (GPSR) is a set of rules that requires consumer products to be safe and our obligations to ensure this.

If you have any concerns about our products, you can contact us on

ProductSafety@springernature.com

In case Publisher is established outside the EU, the EU authorized representative is:

Springer Nature Customer Service Center GmbH
Europaplatz 3
69115 Heidelberg, Germany